BRITISH
ELECTRO-OPTICS

BRITISH ELECTRO-OPTICS

PROCEEDINGS OF A CONFERENCE
AT THE BRITISH EXPORT MARKETING CENTRE, TOKYO
IN DECEMBER 1975

Edited by

L. R. BAKER

SIRA Ltd
Chislehurst, UK

TAYLOR & FRANCIS LTD

10–14 Macklin Street London WC2B 5NF
1977

First published 1977 by Taylor & Francis Ltd
10–14 Macklin Street, London WC2B 5NF

ISBN 0 85066 101 3

Printed and bound in Great Britain by
Taylor & Francis (Printers) Ltd,
Rankine Road, Basingstoke,
Hampshire RG24 0PR.

Cover photograph courtesy of SIRA Ltd.

Preface

This book is the outcome of an exhibition of electro-optical components and systems held at the British Export Marketing Centre (BEMC) in Tokyo as a joint venture with the Department of Trade and sponsored by Sira Institute Ltd. The Japanese guest of honour on this occasion was Professor Hiroshi Yoshinaga, Emeritus Professor of Osaka University and Chairman of the Japan Society of Applied Physics. Over the same period, in an adjacent room, and before an invited audience, a group of UK speakers closely associated with important developments in this field, described the background to some of the exhibits and also attempted to illustrate the current status in the UK of this rapidly expanding technology. The official proceedings of this seminar was of limited circulation and published in three sections in Japanese and distributed by the BEMC. This book has been prepared in response to a request for a volume of the same papers available to a wider English-speaking readership. The lectures, which were organised into three sessions, were chosen by a papers committee to encompass important recent developments in detectors, lasers and equipment. In order to set the scene and to present a more complete picture of the current UK work the main topics were preceded where appropriate by a review of the particular subject.

The success of electro-optics owes a great deal to the technical impetus provided by investment in communications—particularly of course in TV—but the efficient conversion of photons into electronic signals is vital to the success of virtually all electro-optical systems. The significant developments in radiation and image detection made in UK industrial and government research laboratories in recent years was appropriately the subject of the first four contributions.

The central group of papers were devoted to important practical applications of lasers. Once again it was the development of the electro-optical component, in this case the source, that enabled the development of new systems. The attention of UK workers has been attracted away from the application of the coherence property of the laser, as required in holography, towards its high energy concentration capability, needed for example in laser machining. Other industrial and military applications are described together with recent lasers capable of frequency tunability over a wide range of wavelengths.

The final group of contributions covered a variety of topics including the use of electro-optical systems in the automatic inspection of industrial products, the development of a range of lenses for forming thermal images capable of use with modern infra-red detectors, the use of electro-optical systems in high speed photography and a borderline paper as far as electro-optics is concerned reviewing the trends in the UK ophthalmic industry.

Inevitably, a collection of seminar papers presented in association with a commercial exhibition of UK equipment cannot present a complete picture of the best current work in the field but hopefully any significant omissions can be included in the next event.

L. R. BAKER

25 October 1976

Contents

Limiting sensitivities of detectors of infra-red and visible radiation*

T. P. McLEAN

Royal Radar Establishment, Malvern, England.

Abstract. The paper surveys the contribution of British scientists to the development of detectors of infra-red and visible radiation and the discovery of the underlying physical effects on which the detectors are based. It then discusses the theoretical limits of detection and how nearly these limits are approached in present practice.

1. Historical introduction

We have been conscious of the existence of visible radiation and at least one way of detecting it, since the origin of mankind. On that timescale, it was only relatively recently in 1800 that the British Astronomer Royal, Sir William Herschel, was the first man to detect and recognize the existence of radiation beyond the red end of the spectrum—infra-red radiation as we know it today. The first infra-red detector used by him was a simple mercury-in-glass thermometer, which he is quoted as describing as of ' exquisite sensibility '. It is of personal interest to me to point out in passing that this thermometer had been made by Alexander Wilson, the Professor of Practical Astronomy at my own parent University of Glasgow. The contributions of these two men are the first of a long series of important contributions made from British laboratories both to the science of the detection of visible and infra-red radiation and to its technological use.

The urge to discover better ways of detecting infra-red radiation has been pursued since Herschel's first experiments. Although it is over the last thirty-five years or so that we have seen the most concentrated activity, it is true to say that, during the intervening one-hundred and forty years most of the basic detection mechanisms had been established. As well as the thermometer, other broad-band thermal detectors like the thermocouple and bolometer had been developed. In addition, the basis for modern photon detectors, having a well-defined long-wavelength cut-off, had been established by the discovery of photo-emission, especially from the surfaces of alkali metals, and of internal photo-effects in materials like the sulphides of thallium and lead. The last thirty-five years, since World War II, has seen the development of ever increasing activity in the exploitation of these mechanisms, in new ways and in new materials, to produce detectors of higher sensitivity, faster speed, and longer-wavelength cut-off. Government research establishments and industrial laboratories in Britain, in association with university workers, have been in the forefront of this work, and I should like to spend a little time highlighting for you some of the important contributions they have made.

Up to the early 1950s, the most extensively developed infra-red photo-conductive detectors were made from polycrystalline films of the lead salts—PbS, PbSe and PbTe. The successful invention and development of the transistor,

---

based on the physics of single-crystal semiconductors, provided the incentive to apply similar techniques to the development of detectors. The difficulties of applying these techniques to the lead salts and the limited near infra-red response of the then conventional semi-conductors, Ge and Si, led to the development of InSb as a semiconductor material and its exploitation in both photoconductive and photovoltaic detectors. These detectors respond out to wavelengths in the 5–$7\,\mu$m range, depending on their temperature of operation. Workers at the Royal Radar Establishment [1] and at the Mullard [2] played a particularly important part in the successful development of the photoconductive version of these detectors. Some years later, in 1965, Putley [3] recognized the importance of free carrier absorption effects at liquid helium temperatures in this same material at much longer wavelengths. The submillimetre detector which he subsequently developed is still one of the most sensitive detectors available of very far infra-red radiation, with wavelengths of several hundred microns and longer.

Success with single-crystal, InSb, photoconductive detectors encouraged the search for other, narrower-gap semiconductors. The incentive here was to achieve significant sensitivity out to around the $10\,\mu$m region of the spectrum, where both a very good atmospheric transmission window exists, and objects at normal ambient temperature have their peak black-body emission. To this day, no simple compound semiconductor has been discovered with a direct, optically allowed, energy gap as narrow as about $0{\cdot}1\,$eV required for detection in this spectral region. However, the practical realization in 1959 by Lawson and his associates [4] at the Royal Radar Establishment that appropriate alloys of cadmium telluride and mercury telluride—known as CMT—could provide detectors with cut-off wavelengths anywhere in this spectral region, and indeed at shorter wavelengths also, has proved to be an extremely significant one. The idea has been exploited on an international scale and has now been developed to the extent that CMT detectors are commercially available with performance figures approaching very closely to the ideally achievable limit. I shall have more to say about that later. More recently, it was realised that similar spectral response could be achieved using appropriate alloys of the tellurides of lead and tin—known as LTT. Detectors made from this alloy can be claimed to have performances rivalling that of CMT detectors and much debate goes on, concerned with the relative merits of the two. In Britain, the Mullard company devote considerable effort to the development of CMT detectors and have them commercially available whilst the Plessey company concentrate on LTT.

For the detection of infra-red radiation of a given wavelength, the utmost sensitivity and speed of response are always achieved by using a semiconductor detector, with an appropriately chosen cut-off wavelength. When this ultimate in performance is not required, the versatility and the lack of the cooling often required by semiconductor detectors, make broad-band, thermal detectors attractive. Some members of this family of detectors were amongst the earliest to be developed—the thermopile and the bolometer. But in 1962, Cooper [5], working at the Imperial College, London, realised that the pyroelectric effect, exhibited by ferroelectrics and some other crystalline materials, could form the basis of another useful type of thermal detector. The effect can be exploited by forming a temperature-dependent reactive element in a circuit; the reactive nature of the process allows it to be used at frequencies above those at which other thermal detectors begin to deteriorate in performance. Putley [6] at the Royal

Radar Establishment realized the significance of this and has played a key role in having these devices developed.   They are now commercially available from the Mullard and Plessey companies with sensitivities and speeds of response better than any other uncooled thermal detectors.   The pyroelectric effect has also been used as the basis for the first television-type camera tube for thermal imaging.

Two other ideas of British origin which have been developed for novel forms of detectors are worthy of mention.   The photon-drag detector uses the momentum imparted by an absorbed photon to the absorbing carriers in a semiconductor. This can be used to generate a voltage proportional to the intensity of a uni-directional beam of radiation.   Gibson and Kimmit [7] at Essex University have used this idea to develop an extremely fast and robust detector using Ge at room temperature.   The responsivity is low and so it is not suitable where high sensitivity is required.   However it is eminently suitable for the study of fast pulses from CO_2 lasers.

The effect, discovered by and named after Brian Josephson of the Cavendish Laboratory, and for which he received the Nobel Prize last year, can be put to good effect in detecting radiation.   Detectors using this effect respond to photons with energies comparable to superconducting energy gaps.   They can be used therefore for sub-millimetre wave radiation and prove to be extremely fast and sensitive.

Finally, it is worthy of note that, in addition to these contributions to the development of infra-red detectors, the original ideas, on which modern television is based, were those of a Scotsman, J. L. Baird, in the 1920s.   Significant contributions were later made to the development of television camera tubes by McGee [8] and his colleagues working for EMI Electronics.

I hope that, with this short account, I have reminded you of the outstanding contribution which scientists in Britain have made over the years to the problems of the detection and exploitation of visible and infra-red radiation.

2.   Detection limits

Before going on to describe the performances available from present day detectors, I should like to recall to you what the fundamental limiting performances are which, in ideal circumstances, one could hope to achieve.

The ultimate sensitivity achievable by any radiation detector is determined by the nature of the fundamental interaction between the radiation field and the detector material.   The nature of this interaction is such that the individual photo-events, which initiate the detection process, are related to the exciting radiation only in a statistical, rather than a deterministic way.   Thus, although exciting radiation of constant intensity will give rise to photo-events at a correspondingly constant mean rate, the instantaneous rate will fluctuate randomly about this mean.   It is this fluctuation—be it associated with the signal radiation to be detected or with the general radiation field of the environment in which the detector finds itself—which ultimately limits the sensitivity we can expect to achieve with any detector.

The limitations produced by these fluctuations reveal themselves in a variety of ways in different circumstances.   First of all consider the case when the only radiation to which the detector is exposed is that which we wish to detect.   Then the ideal limit of the smallest incident power which can be detected is set by having a noise-equivalent-power (NEP) which ensures the production on average of one photo-event per detector sampling time. Then in this noise-in-signal limited situation: [9]

$$NEP = \frac{2hc}{\lambda \eta} B \tag{1}$$

where hc/λ is the photon energy of the radiation of wavelength λ, η is the detector quantum efficiency and B is the bandwidth of its associated electronics. Modern photomultiplier tubes with cut-off wavelengths in the visible and very near infra-red can be made to approach this ideal limiting performance very closely.

It is impossible to shield completely detectors with longer cut-off wavelengths from the thermal emission of their surroundings. This so-called background radiation produces a fluctuating detector response against which the real signal to be detected must compete, and whose fluctuations are much larger than those produced by the signal itself. In these circumstances, the NEP is given for a photoconductive detector [9] by

$$NEP = \frac{2kT}{\lambda} \left[\frac{2\pi kT}{h} \eta AB \int_{hc/\lambda kT}^{\infty} dx \frac{x^2 e^x}{(e^x - 1)^2} \right]^{1/2} \tag{2}$$

and is a factor of $\sqrt{2}$ smaller for photovoltaic devices due to the absence of re-combination noise. This expression is for the NEP at the cut-off wavelength of the detector, and assumes that the detector sees a background at temperature T, extending over a hemispherical field-of-view.

The wavelength dependence of expressions (1) and (2) is shown in figure 1. These expressions differ in their dependence on detector area A and bandwidth B. As a result, misleading comparisons often arise from using an area of 1 cm² and a bandwidth of 1 Hz. I have tried to be more realistic and used an area of 10^{-4} cm²

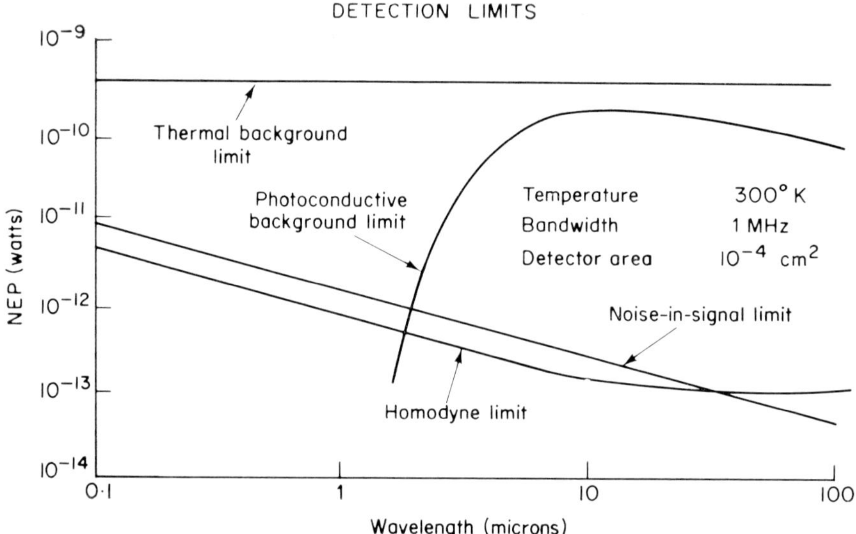

Figure 1. Wavelength dependence of various detection limits.

—that is a square detector of $100\,\mu$m sides—and an operating bandwidth of 1 MHz. The lowest limiting performance of $\approx 3 \times 10^{-10}$ W occurs just beyond $10\,\mu$m and its position there corresponds to the peak of the black-body emission curve at 300K.

Broad-band thermal detectors are even more severely limited in sensitivity by background radiation since, to a first approximation, they respond to radiation at all frequencies. Their noise-equivalent-power consequently turns out to be independent of the operating wavelength; it is given by [9]

$$NEP = \frac{4}{hc}\left[\frac{(\pi kT)^5}{15\eta}AB\right]^{1/2} \tag{3}$$

which for the parameters we have just used works out to be $5 \cdot 5 \times 10^{-10}$ W. This limiting sensitivity is also shown on figure 1.

For any detector, the background limit given by whichever is appropriate from expressions (2) and (3) can be improved on by two methods—the field-of-view can be restricted and the spectral response can be narrowed down as far as the spectral bandwidth of the signal will allow. Although this can be, and indeed is, done to a limited extent, it requires the use of cooled aperture stops and filters. By doing so, one decreases the amount of background radiation which the detector sees and consequently reduces the associated fluctuation noise.

Another technique, which in certain circumstances can be much more effective, is to use homodyning or heterodyning. This effectively swamps any other noise in the detector with that from the local oscillator and at the same time amplifies the effective signal by mixing it with the local oscillator. The detector field-of-view is then restricted without the use of aperture stops, by the fact that only radiation whose wavefront matches that of the local oscillator over the detector mixes with it in a constructive fashion; and the spectral bandwidth of the detector is narrowed by the mixing process, down to that of the electronics. The noise-equivalent-power which can be achieved in this way is controlled by, amongst other things, the coherence area and coherence time of the signal radiation relative to the detector area and sampling time. In the best situation, when coherence area and time are the larger, the NEP which can be achieved is given by

$$NEP = \frac{hc}{\lambda\eta}B \;\; \text{for} \;\; \lambda \ll \lambda_0 = \frac{hc}{kT}$$

$$ \tag{4}$$

$$= 2kTB \;\; \text{for} \;\; \lambda \gg \lambda_0$$

For an uncooled detector $\lambda_0 = 50\,\mu$m and the NEP at long wavelengths is 8×10^{-21} W Hz^{-1}. We see by comparing this expression with (1) that, by homodyning, one can in certain circumstances recover at short wavelengths the performance of a noise-in-signal limited detector. The factor of two difference between (1) and (4) arises from differences in the definition of signal-to-noise ratio and is not a fundamental difference. This, and most other aspects of homodyne detection are discussed in a recent, definitive paper on the subject by Jakeman, Oliver and Pike [10]. The noise-equivalent-power (4) is also shown in figure 1 for comparison purposes.

3.   Current detector sensitivities

The sensitivities achieved in various detectors do not change rapidly with time. It is therefore appropriate to base a discussion of currently achievable sensitivities on figures given in two excellent review articles by Putley in 1966 [11] and 1973 [12]. In figure 2, I reproduce a figure from his earlier article which shows the performance available from a variety of detectors relative to the ideally achievable performance from a background limited photoconductive detector. The main point which I want to bring out here is the relatively good performance achievable near cut-off with the InAs, InSb and doped-Ge detectors. All of these detectors require cooling. This can be a particularly severe limitation on the usefulness of the doped-Ge devices which can call for temperatures as low as liquid helium. Figure 3, which is taken from Putley's later paper, shows the impact which the alloy semiconductors cadmium mercury telluride (CMT) and lead tin telluride (LTT) detectors have had over recent years. These can be made with varying alloy compositions to provide cut-offs over a wide range of wavelengths and can provide performances equal to and better than earlier detectors

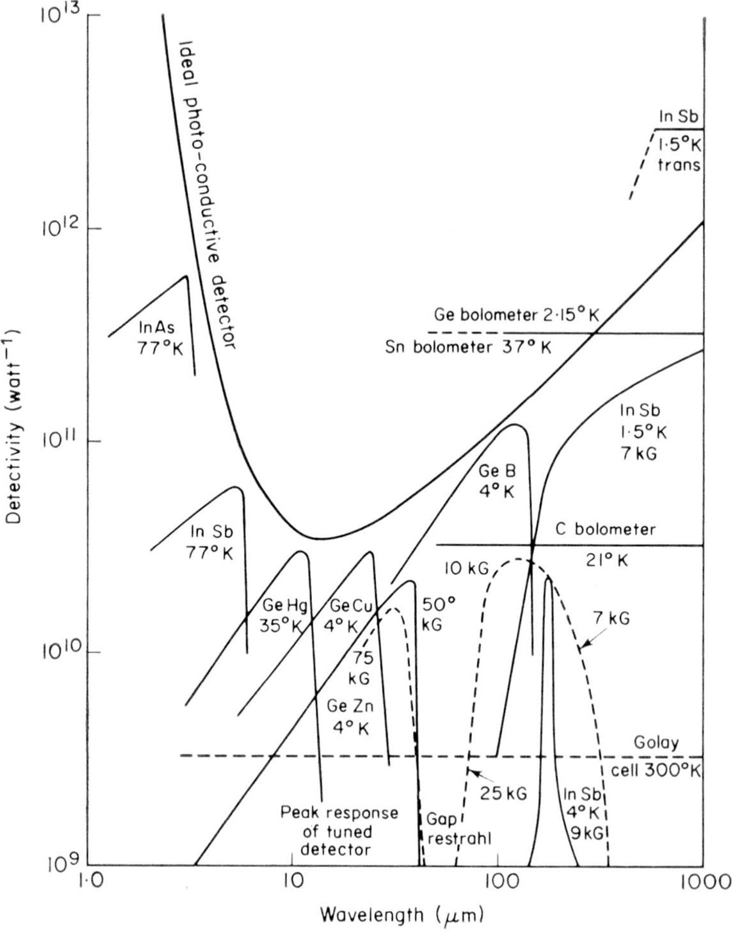

Figure 2.   Performance of far infra-red detectors [11].

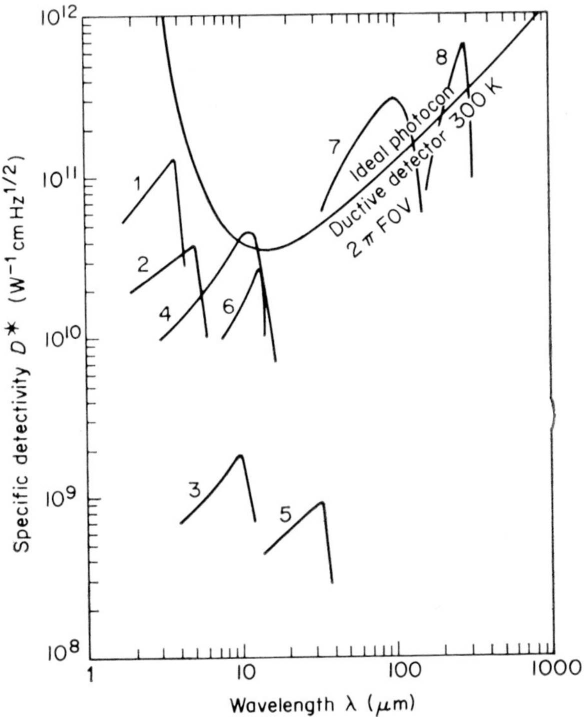

Figure 3.   Spectral response of cooled photoconductors [12].

(1) $Hg_{1-x}Cd_xTe$	$\times 33\%$	Temperature $-80°C$	180° field-of-view
(2) $Hg_{1-x}Cd_xTe$	$\times 27\%$	Temperature $-80°C$	180° field-of-view
(3) $Hg_{1-x}Cd_xTe$	$\times 18\%$	Temperature $-80°C$	180° field-of-view
(4) $Hg_{1-x}Cd_xTe$	$\times 20\%$	Temperature 77 K	60° field-of-view
(5) $Hg_{1-x}Cd_xTe$	$\times 18\%$	Temperature 24 K	24° field-of-view
(6) Pb_xSn_xTe photovoltaic	$\times 80\%$	Temperature 77 K	180° field-of-view
(7) Ga doped Ge		Temperature 4 K	25° field-of-view
(8) Extrinsic GaAs		Temperature 4 K	15° field-of-view

Curves (1), (2) and (3) are based on Chiari and Jervis (1971), curve (4) on the Mullard Infrared Detector Handbook, curve (5) on Saur (1968), curve (6) on Rolls *et al.* (1972), curve (7) on Moore and Shenker (1965) and Jeffers and Johnson (1968) and curve (8) on Stillman *et al.* (1971).

with less cooling; around $10\,\mu m$ cut-offs, liquid nitrogen cooling is sufficient and at shorter wavelengths thermoelectric cooling is adequate.

A general point emerges from this figure which is worth noting. Some years ago when sensitivities were less good, the background limit curve served as a goal which one could strive to reach, without asking too many questions about what it meant in detail. However now that performances are close to, and even sometimes apparently better than this limit, we must be much more careful about what it all means. The ideal background limited performance is a function of the background which the detector is allowed to see by its field-of-view and spectral acceptance bandwidth. Glib statements about detectors being background limited must therefore be treated with caution; one must be sure that the detector can achieve this performance in the environment the user wishes to employ rather than in one chosen by the manufacturer.

Finally in figure 4, which is again taken from Putley's later paper, we see the performances available from a variety of uncooled detectors. In particular, the horizontal lines show what can be obtained from thermal detectors of various types. The attractiveness of the pyroelectric detector is obvious from this, particularly when its robustness compared to its nearest rival, the Golay cell, is borne in mind.

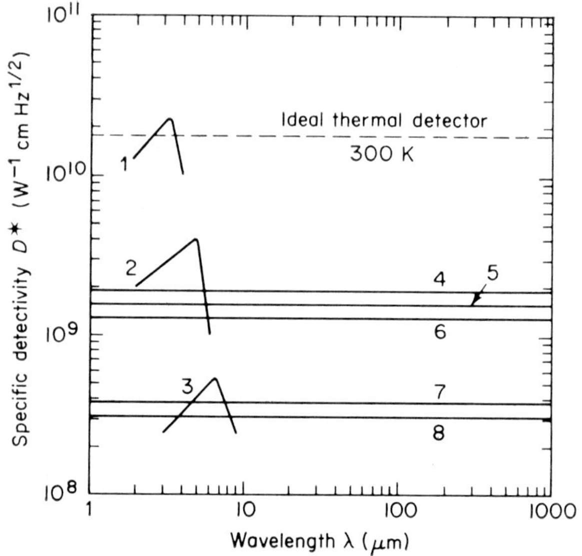

Figure 4.   Performance of some room temperature detectors.

(1) $Hg_{1-x}Cd_xTe$ $\times 35\%$
(2) $Hg_{1-x}Cd_xTe$ $\times 26\%$
(3) $Hg_{1-x}Cd_xTe$ $\times 20\%$
(4) TGS pyroelectric detector
(5) Golay cell
(6) Spectroscopic thermopile
(7) Thermistor bolometer
(8) Thin film thermopile.

Curves (1), (2) and (3) are based on Chiari and Jervis (1971).   Curves (7) and (8) are based on the Barnes engineering detector brochure.   Curves (4) to (8) are idealized in that they assume perfect transmission through the detector window, and uniform absorption by the detector itself.   In practice the spectral transmission of the window will determine the spectral response obtainable and in addition the performance of thermopiles and bolometers tends to fall off beyond about 20 μm due to incomplete absorption by the blackening of the detector.

The ascendancy of the pyroelectric detector over other thermal detectors and its general versatility in both speed and sensitivity is perhaps one of the more outstanding developments of the last few years.   There is still room for improvement as can be seen from figure 4 and ideas are around for how this could be achieved.   However achievement of the ideal background limited sensitivity is extremely unlikely.   For this assumes that the detector is thermally coupled to its surroundings through only the radiation field.   To eliminate all convective and conductive coupling will not be easy.

Many of the applications of infra-red detectors are concerned with imaging and involve the use of scanned arrays of detectors. It is then not sufficient to have a few detectors of high performance. One must be able to produce many detectors, all of high performance and in close proximity to one another. In addition, great stress is laid on either uniformity of detector performance over the array or fast detector response depending on the mode of scanning used. Considerable advances have been made in these problems in recent years leading to the development of impressive imaging systems.

Finally, having introduced the subject of imaging, I should like to mention the development of photocathodes for imaging in the near infra-red. Trialkali photocathodes in the form of S20s and S25s have been improved slowly over the years since their first invention by Sommer in 1955. They are sensitive in the visible and very near infra-red. The discovery in 1965 by Scheer and van Laar of how to produce on GaAs a surface with negative electron affinity opened up the way to more efficient photocathodes, extending further into the infra-red. This can be achieved using not only GaAs but probably even more effectively with ternary or quaternary III-V compounds with lower energy gaps or with Si. Although photocathodes of this type are not yet generally available, interesting progress with their development is reported. They should be seen in much more general use in the next few years.

This has been a very rapid summary of the current detector scene. In the following papers, my colleagues will discuss certain aspects of it in more detail.

4. References

[1] AVERY, D. G., GOODWIN, D. W., and RENNIE, A. E., 1957, *J. Sci. Instr.*, **34**, 394. HULME, K. F., and MULLIN, J. B., 1962, *Solid State Electronics*, **5**, 211.
[2] MORTEN, F. D., and KING, R. E. J., 1965, *App. Optics*, **4**, 659.
[3] PUTLEY, E. H., 1965, *App. Optics*, **4**, 649.
[4] LAWSON, W. D., NEILSON, S., PUTLEY, E. H., and YOUNG, A. S., *J. Phys. Chem. Solids*, **9**, 325.
[5] COOPER, J., 1962, *J. Sci. Instr.*, **33**, 467.
[6] PUTLEY, E. H., 1970, in *Semiconductors and Semimetals* (edited by Willardson and Beer), (Academic Press). **5**, 259.
[7] GIBSON, A. F., KIMMIT, M. F., and WALKER, A. C., 1970, *App. Phys. Letters*, **17**, 75.
[8] McGEE, J. D., 1950, *Proc. I.R.E.*, **38**, 596; 1950, *Proc. I.E.E.*, **97**, 377.
[9] KRUSE, McGLAUCHLIN, and McQUISTAN, 1962, *Elements of Infra-red Technology*, (Wiley).
[10] JAKEMAN, E., OLIVER, C. J., and PIKE, E. R., 1975, *Adv. in Physics*, **24**, 349.
[11] PUTLEY, E. H., 1966, *J. Sci. Instr.*, **43**, 857.
[12] PUTLEY, E. H., 1973, *Phys. in Technology*, **4**, 202.

Photon detectors

E. M. WORSTER

Electron Tube Division, EMI, 243 Blyth Road, Hayes, Middlesex, England

Abstract. Various problems arise in the detection of light of differing wavelengths and intensities. Detectors fall roughly into two categories, namely scanned and non-scanned devices. The scanned variety includes television camera tubes whereas the non-scanned category comprises photomultiplier tubes, vacuum and solid-state photodiodes and image intensifiers.

Image intensifiers and television camera tubes such as vidicons have the advantage that they can provide a bright image of the object being observed. In the case of image intensifiers excellent results can be obtained both in terms of signal-to-noise ratio and resolution with very low light levels. The limiting levels in the case of image intensifiers are considerably lower than those associated with vidicons.

It is sometimes more convenient to use an electrical signal from a photomultiplier than the image from a camera pick-up tube or image intensifier. Photomultipliers have the advantages of wide spectral coverage, high gain, high frequency response and very low noise amplification. The pulse of current from the tube output can be used in a variety of ways and the pulse shape and amplitude gives detailed information about the event which produced it.

Applications employing photomultipliers include astronomy, nuclear radiation detection, television, process control, pollution monitoring and photometry.

Image intensifiers have an advantage where an image is essential and EMI tubes have been used in high quality spectroscopy, photography, crystallography, microscopy and non-destructive testing.

At long wavelengths one often has to use solid-state detectors such as photodiodes. The photoconductive cell is only useful in applications where a slow rise time and a high temperature sensitivity are not critical. Silicon photodiodes have better response times and sensitivity well into the infra-red. Avalanche photodiodes have these advantages plus amplification factors of over 200 times and are often used in distance-measuring equipment and optical fibre communications.

1. Introduction

This paper considers the means of detecting light of various wavelengths and intensities and applications of these detectors. It also touches upon the advantages and disadvantages of the various devices.

The spectrum under consideration stretches from very short wavelengths, through the ultra-violet and visible regions into the near infra-red. The detectors available for this sort of spectral coverage can be roughly divided into two categories:
  1. Scanned devices
  2. Non-scanned devices

2. Scanned detectors

Television camera pick-up tubes are the best known of the scanned photodetectors. The vidicon is the most common camera tube and briefly consists of a tubular glass envelope (figure 1) of about 1 in. diameter and 6 in. length and

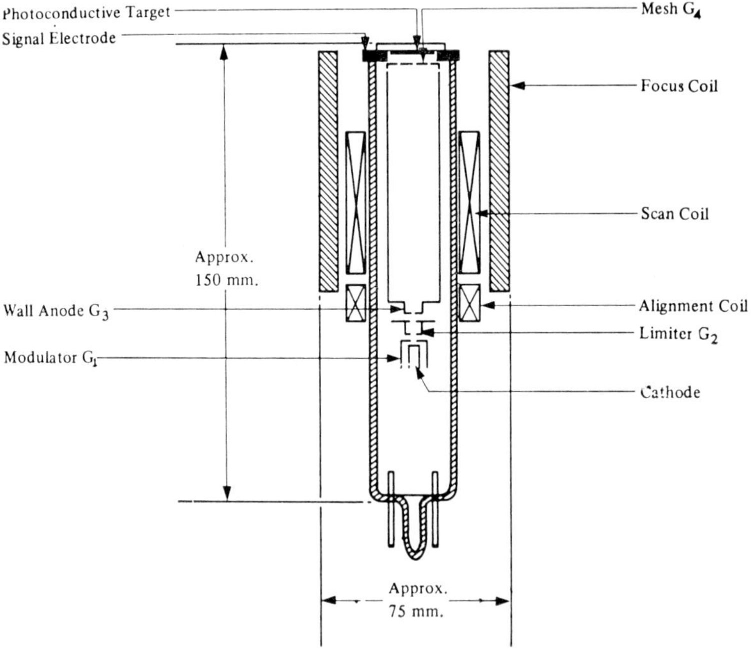

Figure 1

has a flat window sealed to it. The window carries on the inside a transparent conductive coating, acting as the signal plate. A layer of photoconductive material is applied to the signal plate and this constitutes the sensitive target. At the other end of the tube is an electron gun which generates the scanning beam. A mesh called the wall anode is mounted on top of the last gun electrode. This mesh serves to produce a strong, uniform decelerating field in front of the target. The beam of slow electrons stabilizes the surface of the target at the potential of the cathode of the gun. Thus, by applying a positive voltage to the signal plate, a potential difference is set up across the target. Where light falls on the target, a more conductive path is established; a current flows across the target and the potential of the illuminated target elements rises towards the signal plate potential. When the beam scans over the target surface, it restores the elements in turn to cathode potential and thereby a train of current pulses is generated in the signal resistor which constitutes the video signal. EMI vidicons employ separate mesh construction, developed by EMI Central Research Laboratories, to give improved resolution. An antimony trisulphide target is used in standard tubes. They are available in a variety of grades for uses ranging from broadcast TV, through medical and industrial applications to simple CCTV installations for general security such as store surveillance.

There is a range of rugged tubes for use under severe environmental conditions. We also provide $\frac{2}{3}$ in. and $\frac{1}{2}$ in. types for miniature cameras developed for internal pipe inspection. The $\frac{2}{3}$ in. tube is gradually replacing the 1 in. tube for some industrial applications.

We have also developed an all-electrostatic vidicon (figure 2) which uses two planar systems at right angles instead of the single rotationally symmetric lens for focusing.

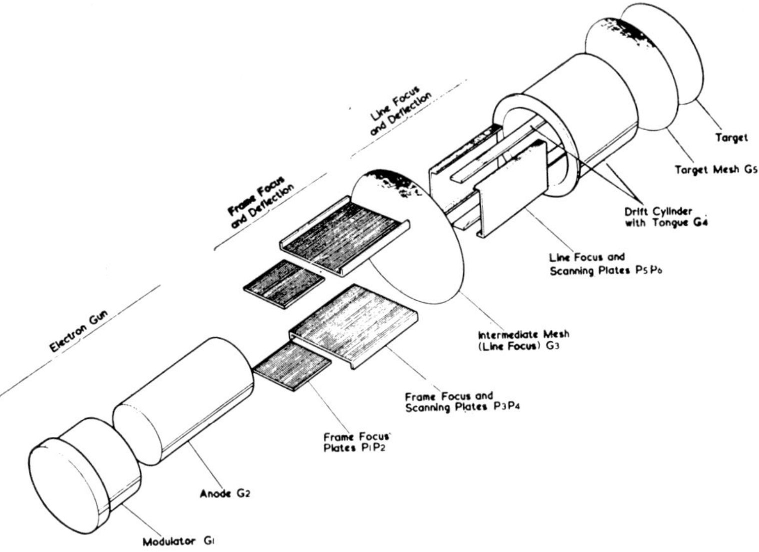

Figure 2

Figure 3

Ultra-violet sensitivity can be obtained by the use of quartz windows and arsenic-doped selenium target layers. The advantage of this tube is that it enables microscopy to be carried out as easily with UV as with visible light. Additionally, since the dark current of this layer is very low, the tube can be used for integration of weak signals over a period of time.

Other types of vidicon include the Plumbicon, a Philips camera tube with a lead-oxide photoconductive layer, tubes with cadmium-selenide layers, silicon targets, and vidicons with intensifier sections mounted in front of the target to give enhanced sensitivity. Fibre-optic faceplate tubes are available for coupling to image intensifiers but it is often preferable to have the intensifier section as part of the camera tube, as in the case of the EMI Ebitron (figure 3).

By using a small vidicon gun this tube can be made to occupy the same diameter as the conventional 1 in. vidicon and its scanning coils. Thus a standard 1 in. vidicon camera can easily be modified to take the Ebitron to extend its low light-level capability. The Ebitron will operate at light levels down to bright starlight conditions (5×10^{-10} W).

For completeness it should be mentioned that new cameras employing self-scanned arrays of photodiodes are now available and are able to supplant vacuum-tube cameras in a number of applications. One should also mention the image orthicon which, due to the advances in vidicon type tube technology, has become obsolescent.

Non-scanned detectors

The non-scanned detector category comprises vacuum and solid-state photodiodes, image intensifiers and photomultiplier tubes.

Photomultiplier tubes (PMTs) can be used in many applications and in most of them are employed to measure light of a very low level, down to a few photons/sec. Visible light (figure 4) which covers the wavelength range of approximately 400–700 nm does not comprise the total region of interest. The shorter wavelengths in the UV and to some extent the longer IR wavelengths are also measured by PMTs.

A vacuum photodiode consists of a photocathode which converts incident light into photoelectrons and an anode which collects these electrons. A small current therefore flows through the diode and this can be externally amplified by conventional means. It will be proportional to the intensity of the incident light.

A PMT has secondary emitting electrodes (dynodes) interposed between the cathode and anode so that the number of electrons reaching the anode will be many times greater for the same incident light level than is the case with photodiodes. Thus PMTs are effectively photocells with built-in current amplifiers. Their advantage over devices without internal amplification is that very high current gain, of the order of 10^8, can be achieved without significantly degrading the signal-to-noise ratio. Figure 5 shows a range of PMTs.

The spectral range over which light can be detected depends on the photocathode surface material used in the PMT (figure 6). For UV and blue light applications the K_2CsSb 'bialkali' surface is preferred because of its high sensitivity and low dark current. For applications involving red light such as HeNe or ruby laser radiation, the Na_2KSbCs trialkali surface is best of the

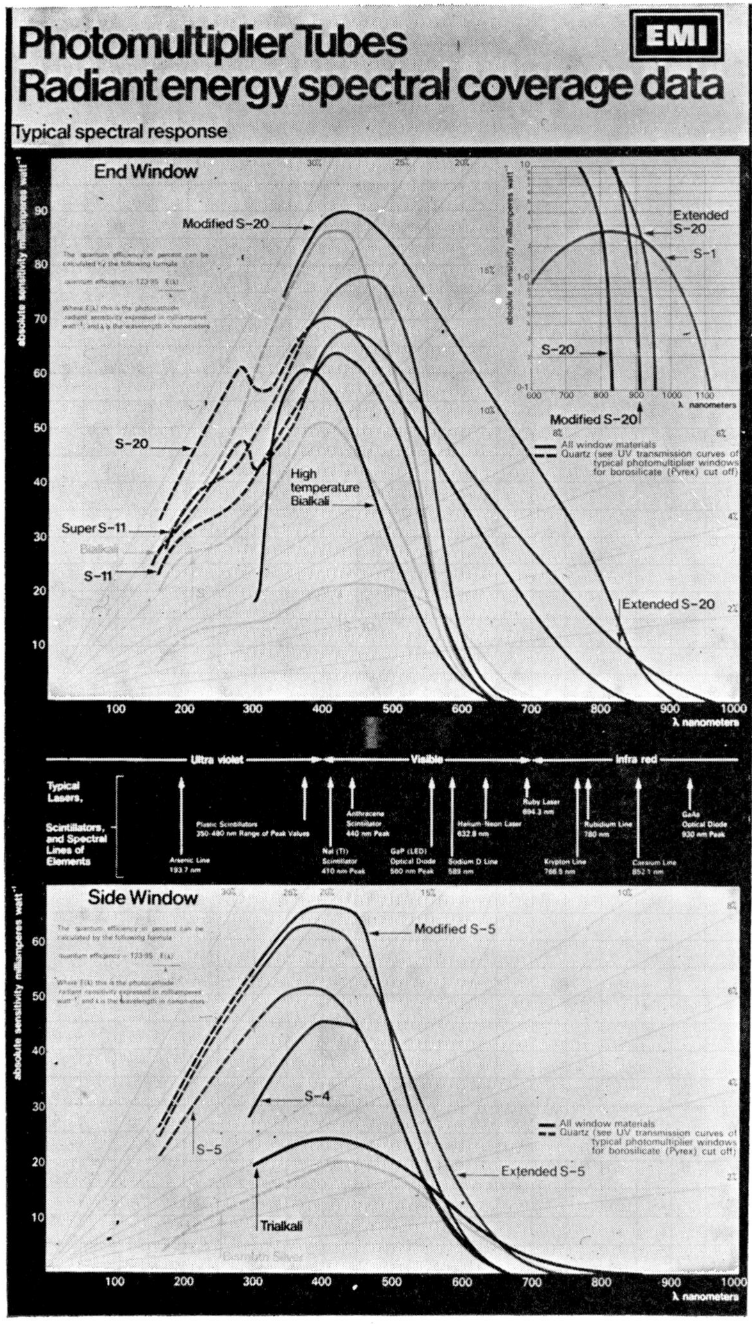

Figure 4

semi-transparent photocathodes. The long-wavelength threshold of the trialkali surface can be extended by modifying the substrate on which it is deposited. In the case of the EMI type 9658B the substrate is the inner surface of the end window and by making this surface a matrix of prisms (figures 7 and 8) the light normal to the outer surface of the window enters the photocathode at a

 E. M. Worster

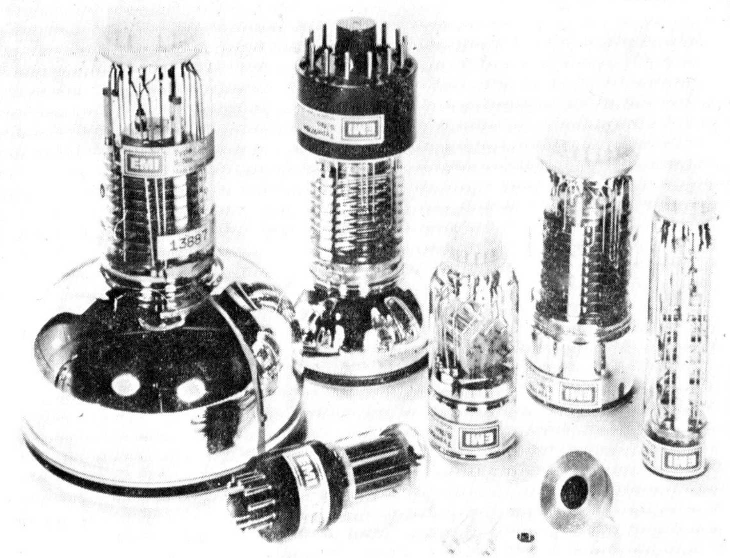

Figure 5

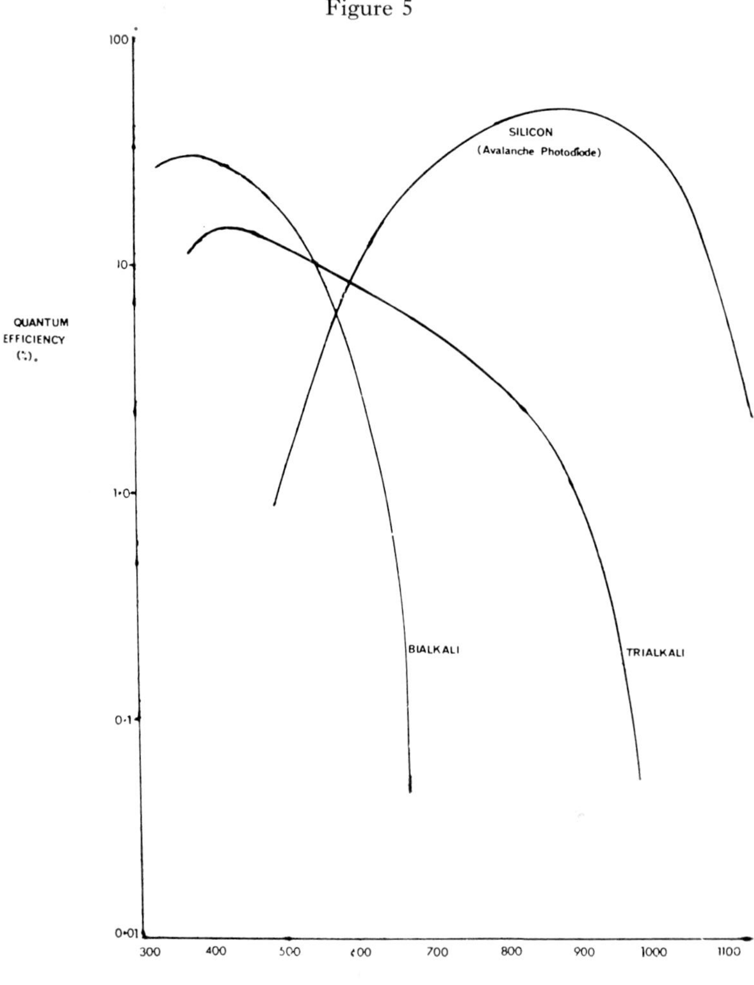

Figure 6. Absolute spectral response of bialkali and trialkali semi transparent photocathodes and silicon photosurface.

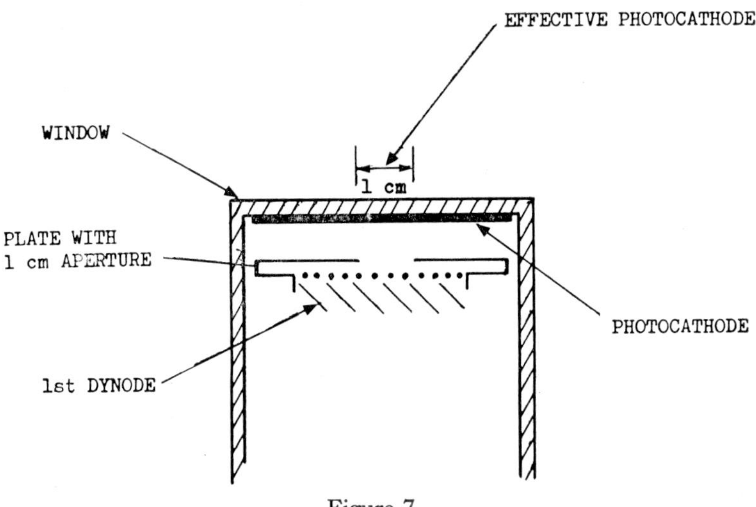

Figure 7

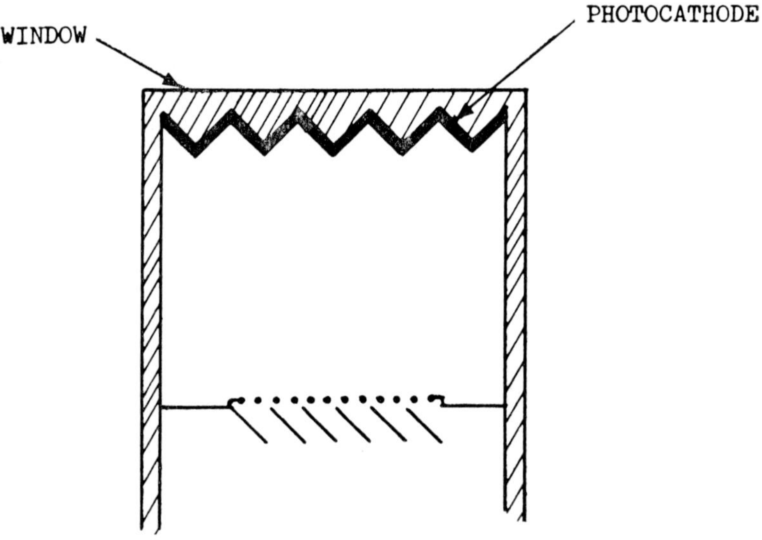

Figure 8

more shallow angle, thereby increasing the absorption path length and the probability of electron emission.

In recent years compounds made from materials such as gallium arsenide (from groups III and V of the periodic table) have been developed which, after treatment with a thin layer of Cs, show negative electron affinity. Such materials give high values of sensitivity right up to the long wavelength threshold but so far these cathodes are only available in opaque form.

The longest wavelength of response commercially available in a PMT is provided by the AgO–Cs (S1) photocathode which extends to approximately 1200 nm but this cathode has a very high dark current associated with it. Also the peak sensitivity is only about 0·4% quantum efficiency at 800 nm.

E. M. Worster

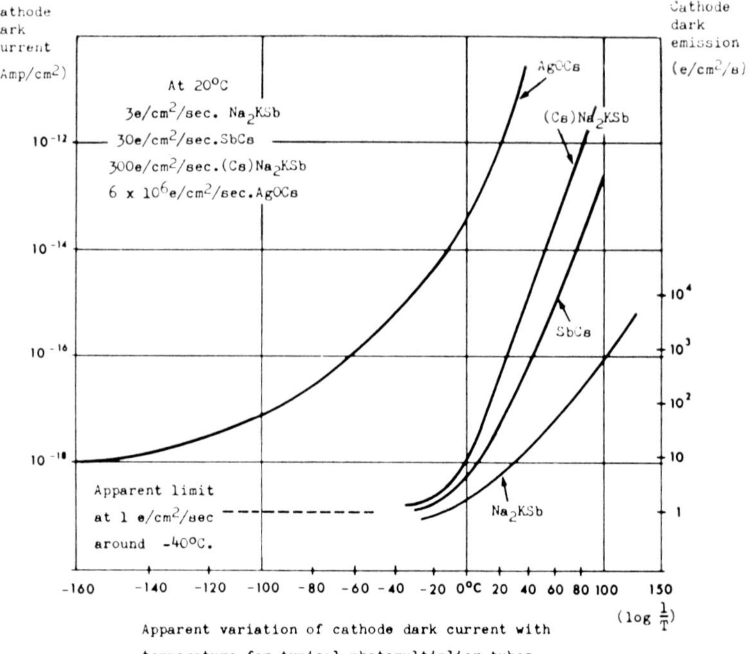

Apparent variation of cathode dark current with
temperature for typical photomultiplier tubes.

Figure 9

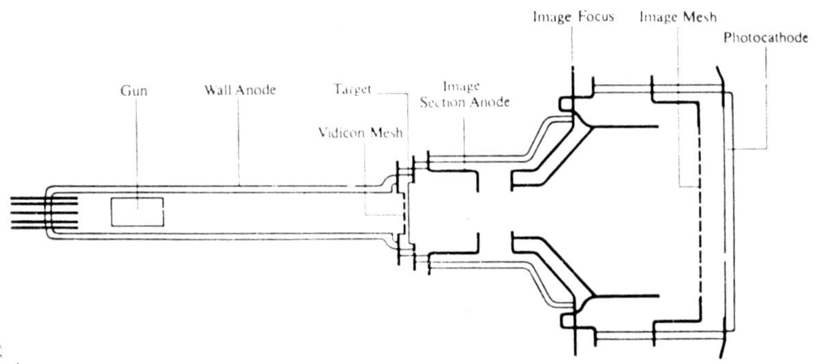

Figure 10

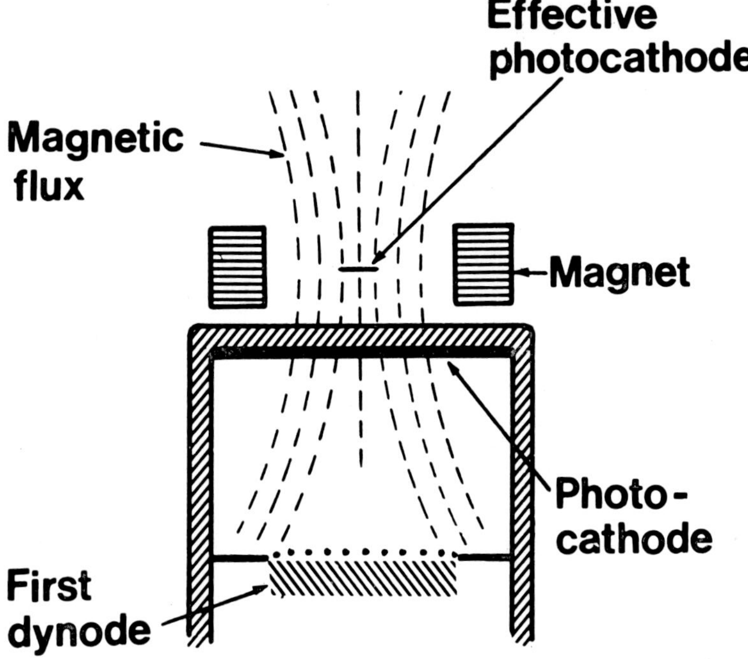

Figure 11

The choice of photocathode for a particular application will be influenced by its spectral response and also by its dark current. The dark current is the current which flows in a PMT in the absence of a light signal and it has been shown that about 90% of this dark current is due to thermionic emission from the cathode area. The lowest dark current at room temperature is associated with the Na_2KSb cathode which is approximately 3 electrons sec^{-1} cm^{-2}, rising through 10 for K_2CsSb, 30 for Cs_3Sb, 300 for Na_2KSb–Cs up to 6×10^6 electrons sec^{-1} cm^{-2} for the AgO–Cs cathode.

Since dark current is predominantly thermal in origin, reducing the temperature at which the tube operates will reduce the dark current. Figure 9 shows the effect of temperature on the various dark emission rates.

Dark current is also proportional to cathode area and restriction of the effective cathode diameter by a proximity focus arrangement (figure 10) or an electrostatic imaging arrangement, as in an image dissector, can reduce dark current considerably.

The effective cathode diameter may also be reduced by the use of a magnetic lens to cause photoelectrons from the cathode to be directed away from the first dynode, except for those produced in the small central area (figure 11). This lens can take the form of an annular permanent magnet or an electromagnet in the shape of a flat pancake coil. The electromagnet has the advantage that the field is adjustable by the variation of current through the coil but at appreciable current levels the local heating effect can partially offset the gain in dark current reduction.

Dark current limits are set not by thermionic emission but by scintillations produced in the window by internal and external radioactivity. The lowest value is obtained with quartz windows.

Cooling tubes not only reduce dark current but can shift the spectral response of a photocathode by reducing the red threshold wavelength and moving the peak response towards the UV, in some cases as much as 10 nm. Temperature variation will also cause a change in gain of approximately -0.5% K^{-1}. This figure is typical and will vary from tube to tube and also from one type to another.

The gain of a PMT is adjusted by the alteration of the interdynode voltage down the tube. Because of the steep gain-versus-voltage characteristic of these devices the power supply from which the dynode potentials are derived must be highly stable. A 1% change in overall voltage can produce a 10% change in gain and users have to be aware that PMTs are not inherently stable, particularly those containing caesium.

The actual passage of current through a tube also disturbs the caesium distribution and thus the gain. It is therefore desirable to stabilize a tube by operating at the appropriate current for a few minutes before making measurements. In any case it is good practice to keep the average anode current to a low level ($\approx 1\ \mu$A for the best stability. This is particularly important in tubes with GaAs–CsO cathodes.

At the anode end of the tube, however, the amplified current may be large enough for space charge to degrade the linearity and the interdynode potential must be high enough to take account of the peak current in use. A less fundamental problem may often cause problems with non-linearity when the distribution of dynode potentials is disturbed by the current drawn from the tube anode. It is necessary to ensure that the dynode divider chain current is high compared with the tube current and the last few stages are adequately decoupled.

One other characteristic of importance is the time response and the cathode to first dynode geometry has a significant influence on this parameter. It is necessary to have high fields and a suitable electron-optical system here to minimize the spread in transit time of electrons from the photocathode into the first dynode. The high field of the linear focused system gives shorter transit time and less spread than the weaker fields in the venetian-blind and box-and-grid structures. Figure 12 shows the various dynode systems.

The rise time of the anode current pulse due to a short light pulse incident on the photocathode varies from 2 ns in the linear focused tubes to 10 ns in the slower types. The fastest tube easily available is the compact focused tube (figure 12*b*) aged so that it will operate at high voltages. This gives a rise time of less than 1 ns.

To summarize, the PMT consists of a photocathode from which photoelectrons are drawn into a secondary emission multiplier where they are multiplied and collected on an anode. The overall gain of the device is adjustable from 10^3 to 10^8 depending on the tube type and can be used in applications where bandwidths of up to 100 MHz are needed.

The applications of these devices are numerous but the best known fields are nuclear radiation detectors employing crystals, and liquid phosphors, spectroscopy, flying-spot scanners in broadcast TV and astronomy.

Short wavelength (UV) radiation is detected principally by the use of electron multipliers in which the radiation is incident on the first dynode which is made from BeCu or AgMgO. Electron multipliers may also be used as ion detectors,

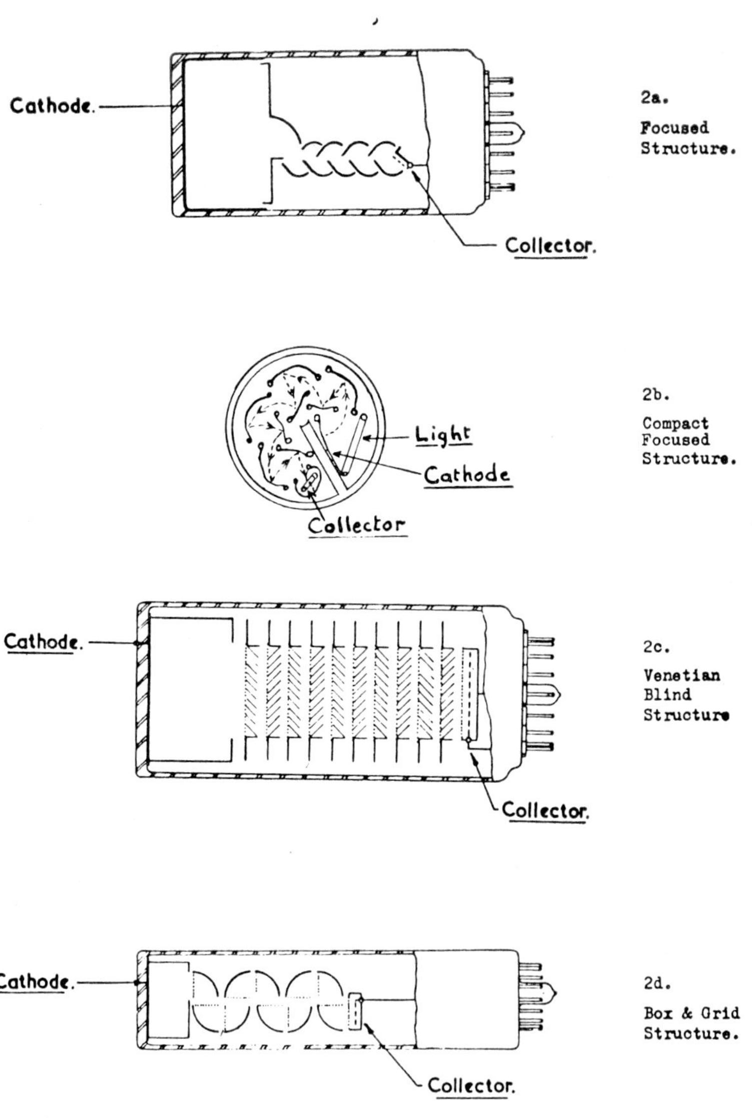

Figure 12

as in mass spectrometers. These multipliers can have between 12 and 20 stages and often incorporate the dynode resistor network. Quadrupole mass spectrometers are extremely sensitive and are used in the detection of toxic atmospheric pollutants such as vinyl chloride monomer.

In nuclear radiation detectors, the gamma ray scintillation counter demands a high photocathode conversion efficiency of light to photoelectrons and a high first dynode gain for a given amount of energy dissipated in the crystal, in order to obtain an accurate measure of this energy. For the measurement of low-energy beta particles such as those from tritium, liquid scintillation counters

                    E. M. Worster

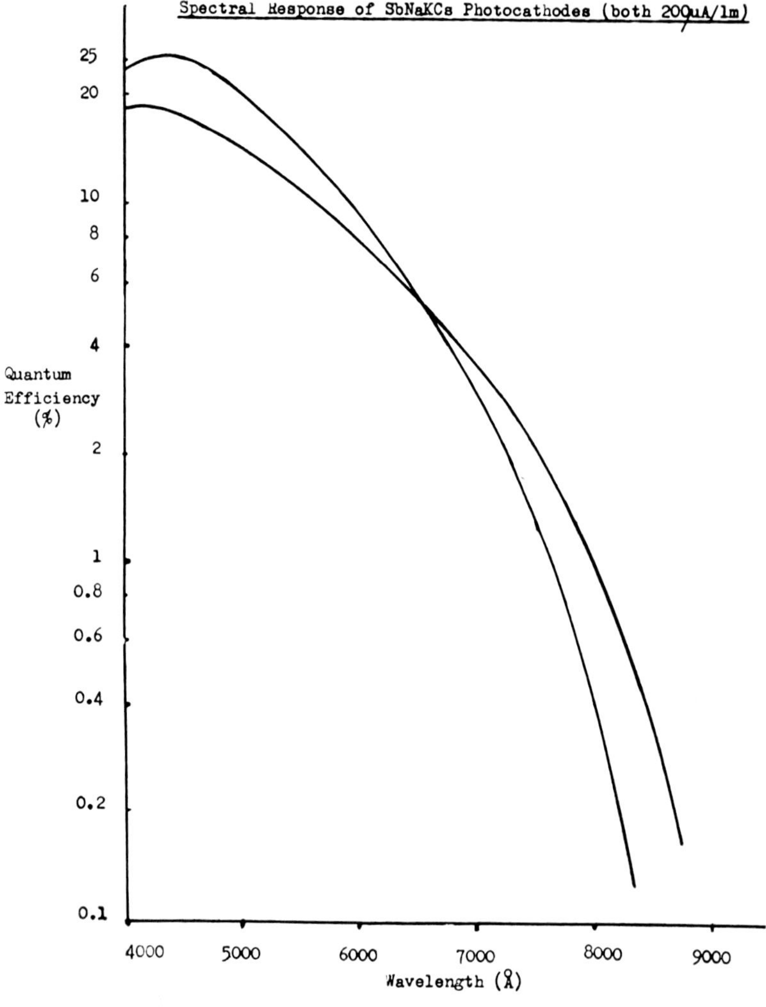

Figure 13

are used with the active material dispersed in the solvent. In this case the scintillator is observed by two tubes working in coincidence at the single electron level, thus eliminating all but the chance counting of simultaneous dark pulses. The average energy of 5 keV only produces enough light in the scintillator to produce 2 or 3 photoelectrons in each tube, so there would be great interference from the dark current and window scintillation if a single tube were used. This application calls for tubes with high blue sensitivity and low dark current: bialkali cathodes are used.

In the field of spectroscopy, the tube cathode must provide a useful quantum efficiency at wavelengths of interest and the trialkali cathode is mostly used. The trialkali cathode can be processed to give the required performance (figure 13).

It is a fact that if monochromatic light is scattered by any material, the scattered light will consist of the incident light wavelength plus ' side bands ' (light of different colours or wavelengths). The difference in wavelength between these side bands and the incident radiation, known as the Raman effect, is like a fingerprint of the sample and is characteristic of the material. Modern laser Raman spectrometers are capable of detecting concentrations of impurity down to one part in 10^5. In this technique the incident light produces pulses corresponding to single photoelectron emission from the photocathode: output pulses within a defined amplitude range, including some background pulses, which are indistinguishable, are recorded in a counter. The advantages arise from the discrimination against leakage components and the ability to integrate the count signal over long time periods to achieve good statistics.

Small concentrations of vaporisable material can be detected by the use of atomic absorption spectrophotometers. These instruments utilise the fact that, when heated, elements absorb light of a precise wavelength according to the particular element concerned. Here, light characterising the material of interest is allowed to pass through a flame in which the sample is burnt and the amount of that light which is absorbed by the flame is a measure of the concentration of the element in the sample. The PMT is used to measure the change in light transmitted through the flame when the sample is introduced.

The characteristic wavelengths which various elements emit when they are excited are used in spectrographs. In the steel industry, a sample of the steel being manufactured is heated to a very high temperature by a spark discharge and the amount of light for the various elements is measured and used to give a precise analysis. In such a unit several PMTs are arranged to accept just one wavelength each, e.g. phosphorus, carbon, to facilitate rapid checking. Since the light emitted by some elements is very small and there is a wide range of colours, the PMTs have to be carefully chosen for each elemental colour.

The choice of PMTs for astronomical observations is obvious when their high sensitivity and low dark current are considered. However, it is sometimes

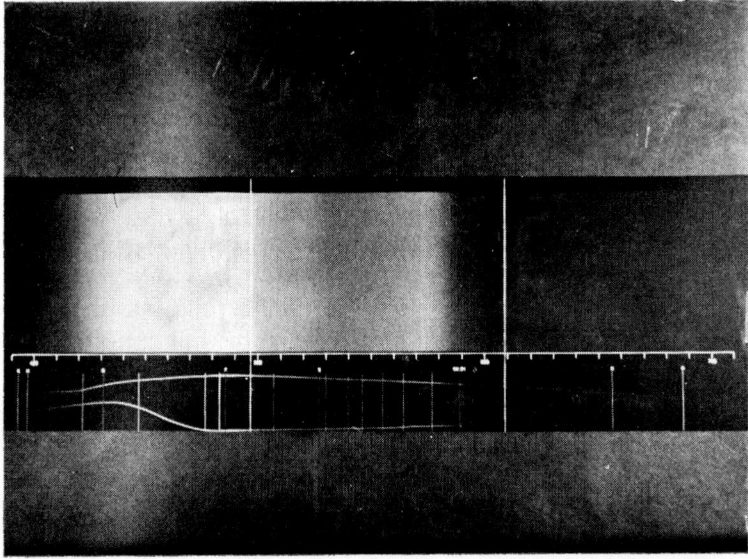

Figure 14

more convenient to observe an image rather than a pulse of current and image intensifiers are being used increasingly in this field.

The EMI range of image intensifiers includes 2-, 3- and 4-stage magnetically focused tubes (figure 14).   They employ the phosphor/photocathode technique of light amplification, which relies on the fact that the scintillation produced in a high efficiency blue phosphor by a 5 keV electron produces approximately 300 photons of blue light of 3 eV energy.   If half these are absorbed in a photo-cathode of 20 per cent quantum efficiency, 30 photoelectrons are emitted.

With 5–10 kV/stage and magnetic field focusing, light gains of about 3×10^5 and a resolution of 40–45 lp/mm can be obtained.   The statistics of the device are very good and one major application is to overcome the reciprocity failure in photographic emulsions to give better detective quantum efficiency.   This is a figure of merit given by the square of the ratio of signal-to-noise at the output to by the signal-to-noise at the input.   Image intensifiers have a value of about 20 per cent whilst photographic plates vary from 0·5 to 5%.

These tubes can be provided with either bialkali (Na_2KSb) or trialkali (Na_2KSb–Cs) photocathodes and with glass, uv-transmitting or fibre-optic windows.   The dark current associated with these tubes is very low, approximately 2 electrons $sec^{-1}cm^{-2}$ in the bialkali, and 200 electrons $sec^{-1}cm^{-2}$ in the trialkali.

The speed of response of this type of intensifier is dependent upon the decay time of the phosphor in the sandwich.   The standard phosphor in the EMI tubes is the P11 but faster phosphors can be used at the expense of some light gain.   In a 4-stage tube with fast phosphors, the light gain is about 100 times lower than that from the equivalent tube with P11 phosphors at the same overall voltage.

Returning to astronomical applications, EMI image intensifiers have been shown to provide very acceptable performance standards in the two principal areas of astronomical application, i.e. ultra high sensitivity spectroscopy, and general high quality spectroscopy and photography.

In the former case the 4-stage tube is applied to enable most of the photons constituting a star spectrum to be counted and positioned within the spectrum with great accuracy.   In the latter case the tube is applied as a general photo-graphic aid on large telescopes, providing maximum image detectivity, i.e. where high sensitivity and detail are both required from a very faint low contrast image.

These applications are obviously more of interest to large telescope operators and tubes for these purposes are being used or are to be supplied to several major observatories (figure 15).

Other applications where the image intensifier has particular advantages include optical, uv, electron and field-ion microscopy, electron and x-ray diffraction (crystallographic investigations), laser research, where the tube can be used in high speed photographic techniques, high energy physics such as viewing streamer chambers and scintillators, high energy radiation in medicine, and non-destructive testing.

Most image intensifiers commercially available operate in the visible region but investigations are being made to provide IR tubes.

It has already been noted that the best photoemissive photocathode for wavelengths beyond 900 nm is the AgO–Cs surface.   Therefore for wavelengths

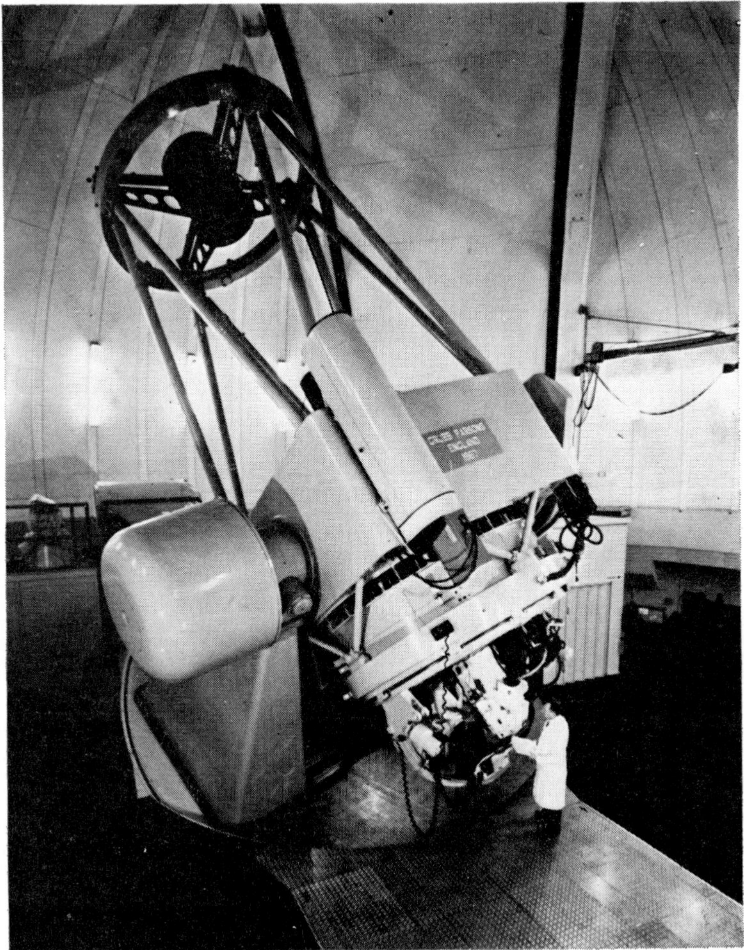

Figure 15

in this region and above, some other form of detector is often necessary and semiconductor devices such as silicon photodiodes and avalanche photodiodes provide a suitable alternative.

The simplest cell of this type is the photoconductive cell which consists essentially of a light sensitive resistor of cadmium sulphide or cadmium selenide.

The resistance of the cell is very high in the dark but falls progressively as the sensitive surface is illuminated. Photoconductive cells are made in various sizes down to units small enough to fit into a TO5 can with a glass window. They are used in electronically operated exposure circuits for obstacle detection and for counting articles on conveyor belts. Disadvantages lie in a high temperature sensitivity and a slow speed of response which is several tens of milliseconds.

Silicon photodiodes have much better time response and a generally more useful spectral sensitivity which extends out to the near infra-red, as shown in figure 16. They consist of p–n junctions and are operated in one or two modes; either in a photovoltaic mode in which the light energy causes the cell to generate

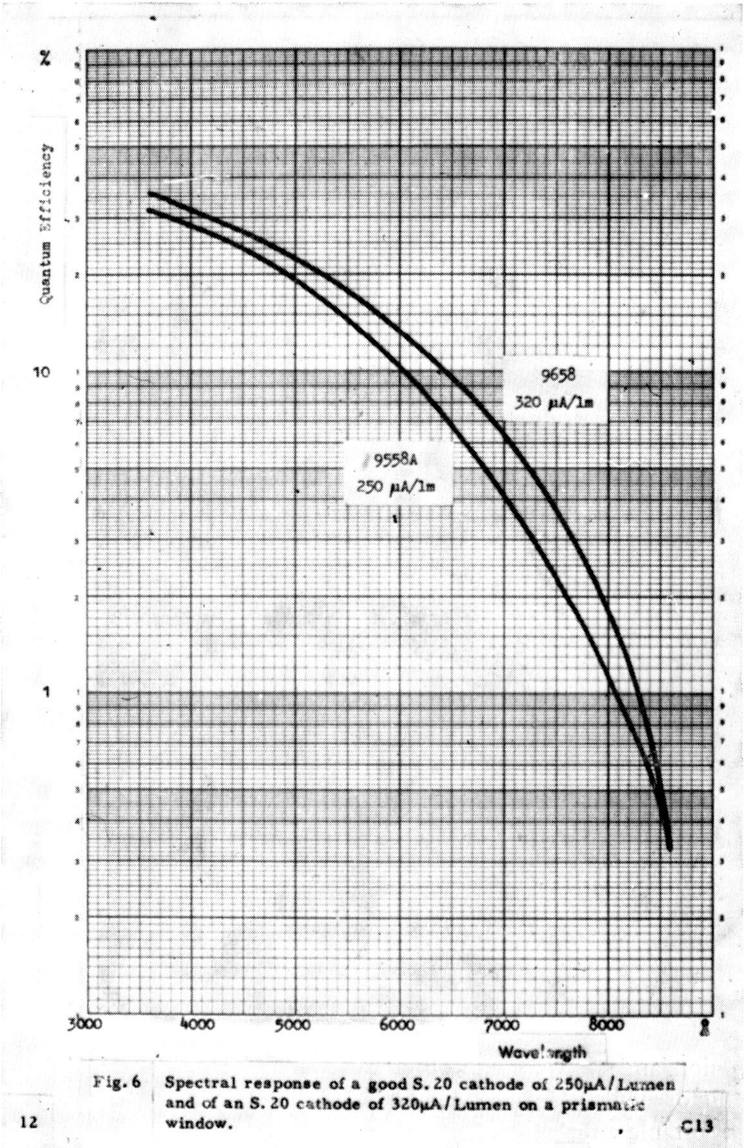

Figure 16

a voltage, or in a reverse-bias mode in which the current is directly proportional
to the illumination on the cell.   The photovoltaic mode is used in solar cells and
the output from the cell has a logarithmic relationship to the level of illumination.
In the reverse-bias mode the limitation on sensitivity is set by the dark current,
which is produced both by leakage and by thermally generated carriers in the
depletion region of the diode.   This current is obviously related to the area of
the cell and units of this kind are made in diameters ranging from about 0·5 mm
to about 2 cm.

The time response in the smaller units is very fast, in the order of 0·5 ns.
Operating with such fast response times, the photocurrent must be amplified by a

Figure 17

wide-band amplifier and this in turn gives rise to noise problems. In order to improve the signal-to-noise ratio, avalanche photodiodes (figure 17) are made in which the bias voltage across the diode is increased to the point where multiplication by collision takes place in the depletion region. This provides a low-noise amplification of the photocurrent, rather analogous to that which takes place in a photomultiplier tube. Amplification factors of 200 times or more are readily achieved with no increase in system noise. A guard ring surrounds the shallow sensitive p–n junction and prevents surface breakdown and the outer edge of the guard ring is buried beneath silicon dioxide which passivates the whole structure. Near infra-red radiation is only weakly absorbed by the silicon and a high responsivity requires a wide depletion range.

In addition to the fast detector response derived from carriers generated by photo absorption in the depletion region, slow rise and decay components are often observed in the output signal. This arises from carriers photogenerated in the bulk silicon surrounding the depletion region which slowly diffuse to, and are collected at, the p–n junction. As a result, the effective d.c. quantum efficiency is generally greater than the pulsed quantum efficiency. A method of substantially reducing the slow component in the guard ring design involves preparing the devices in an epitaxially deposited p-type silicon layer grown on a p^+ type substrate. This ensures a maximum decay time of about 10 ns compared with several microseconds observed in detectors prepared in homogeneous material. For laser-ranging applications the leading edge of the detected reflected pulse is of prime interest and therefore the long tail associated with homogeneous silicon devices is not important. However, in applications where the whole of the pulse is of interest, such as in some laser distance-measuring systems, an epitaxial structure is needed.

Applications in communications are beginning to appear with the increase of interest in the use of glass optical fibres for wide-band communication channels. An advantage of such optical communication systems is that they do not suffer from electromagnetic interference and this is leading to their adoption in aircraft systems. In wide-band communication systems, avalanche photodiodes are particularly applicable.

One of the disadvantages of avalanche photodiodes is that they are very sensitive to changes in ambient temperature and this makes it difficult to maintain a stable operating point. This has been overcome by the use of double chip devices in which two nearly identical chips are mounted in the same can, one masked from light and operated near breakdown to draw a constant current.

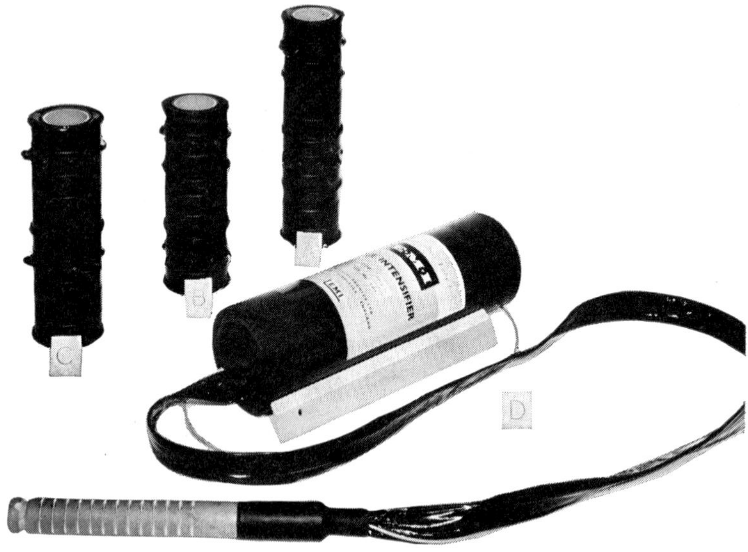

Figure 18

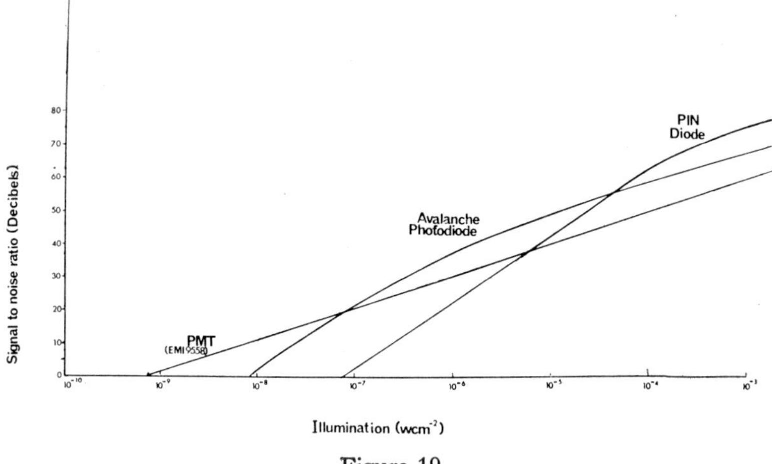

Figure 19

Temperature changes alter the voltage at this diode and the voltage change is used to regulate the bias applied to the active chip.

EMI produces a range of diodes, both in epitaxial and homogeneous silicon and complete photodetector packages are available containing diode, bias controller and pre- and main amplifiers (figure 18). In some cases a d.c.-to-d.c. converter is included enabling the detector to be run from a single low voltage supply.

The foregoing serves to illustrate some of the points to be taken into consideration when choosing a photodetector for a particular application. Some choices are simple, as in the case of whether one wishes to observe an image or a pulse of current. However, in many cases the final decision will depend upon the theoretical signal-to-noise ratio that a particular arrangement will give. Figure 19 shows that for light levels below 10^{-7} W cm^{-2}, the trialkali photomultiplier tube gives the best signal-to-noise ratio whilst at light fluxes between 10^{-7} and 5×10^{-5} W cm^{-2} the avalanche photodiode is preferred. For higher light levels the PIN diode is the best choice from a signal-to-noise point of view.

It has not been possible within the scope of this review to cover all the applications for photodetectors or to explore fully the merits and disadvantages of each type. However, it does demonstrate that a large number of widely differing applications exist in which the measurement of very weak light levels can be significant in increasing our knowledge and in the control of the quality of life.

Image intensifiers and infra-red detectors

M. H. JERVIS and M. J. NEEDHAM

Mullard Ltd, Millbrook Industrial Estate, Southampton

Abstract. This paper summarizes the characteristics of image intensifiers and infra-red detectors that affect image quality and gives typical performance figures for current systems.

1. Introduction

Mullard has had over 15 years' experience in the manufacture of image intensifiers and over 18 years' experience in the manufacture of infra-red detectors. This long experience has made possible the development of the wide range of technologies necessary for the manufacture of these devices. The company manufactures the special glasses and glass fibres for incorporation in the image intensifiers and the crystals required for the fabrication of infra-red detectors.

These types of device require vacuum encapsulations and this has meant that facilities to ensure the necessary cleanliness of components have had to be developed. These now ensure many thousands of hours of operational life.

A critically important component of the image intensifier is the photocathode. Mullard have developed the S25 photocathode which gives an enhanced long-wavelength response out beyond 850 nm. The production of these cathodes takes several hours and has now been computer controlled to ensure consistently high sensitivity and quality.

2. Image intensifier tubes

Over the past 5 years microchannel plates have been developed which enable a new range of intensifiers to be made which are substantially smaller than the cascade tube and which can offer improved resolution, lower distortion and a built-in control for highlight suppression.

A range of image intensifiers has been developed. Emphasis is put not exclusively on one or two characteristics but on the best possible overall performance in each specific field of application.

The techniques for incorporating these plates into intensifier tubes have now been developed to the point where the tubes can be thoroughly aged during manufacture to achieve operational lives in excess of 5000 h.

2.1. *Image intensification*

In the conventional intensifier tube, an intensifier stage consists of a photocathode input surface, an electron-lens/intensifier, and a phosphor viewing screen.

The light that falls on the photocathode triggers the emission of electrons, which are accelerated across a potential of several thousand volts towards the phosphor screen. They are focused onto the screen by deflection plates charged with similar high voltages and, on hitting the screen, are converted back into light at higher intensities. As in optical lenses, the image is inverted in the process.

A single intensifier stage of this kind can achieve a luminous gain of only about 100. Since this is not enough for most night-viewing applications, three such stages are usually coupled together with fibre-optic windows. The resulting assembly of three tubes is known as a cascade image intensifier and is shown in figure 1.

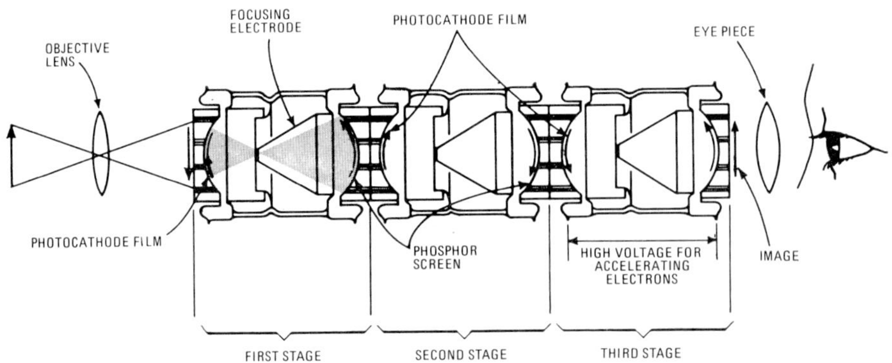

Figure 1. Cascade image intensifier comprises three stages each with a gain of about 100.

Recent developments in intensifier technology, however, make it possible to increase the gain of a single intensifier stage by building in a level of electron multiplication just in front of the phosphor screen. This is the function of the microchannel plate. Because the plate has a variable gain that may exceed 10 000, the resulting tube has an overall gain ranging beyond that of the standard three-stage cascade intensifier.

Two approaches to microchannel intensifier tube design have emerged—the wafer tube and the inverter tube. The main difference between them is in the focusing of the electrons between the photocathode and the channel plate. In the wafer tube (figure 2 (*a*)), so called proximity focusing—in which the close proximity of the two electrodes limits the sideways spread of electrons—is used for electrons both entering and leaving the channel plate. This means the tube can have an overall length of approximately 30 mm, including an inverting fibre-optic plate.

Alternatively, in the inverter tube (figure 2 (*b*)) electrons leaving the photocathode are electrostatically focused on the channel plate (much as electrons in a conventional cascade tube are focused on its phosphor-coated output window), and proximity focusing is used between the channel plate and the output screen. We have concentrated more on the development of the inverter type of microchannel tube because it can combine long tube life with high screen brightness and good image quality.

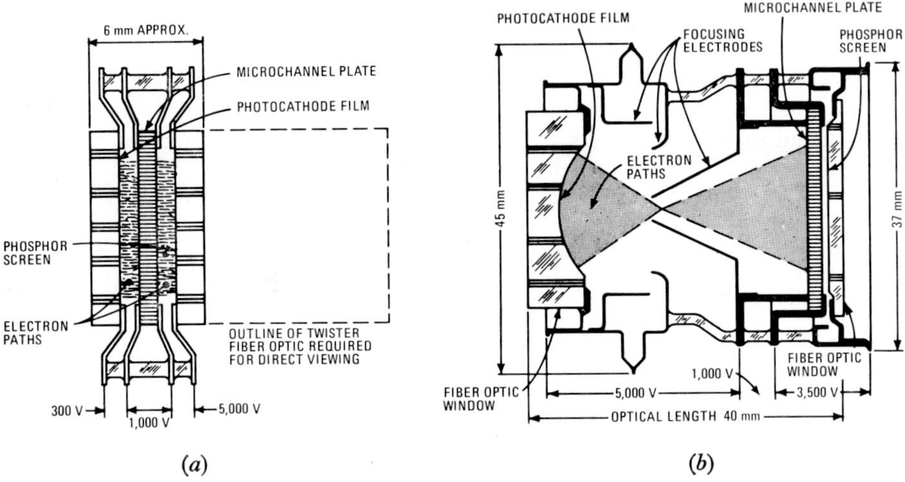

Figure 2.   Microchannel intensifier tubes.   (*a*) Wafer tube.   (*b*) Inverter tube.

2.2. *Inverter tube advantages*

The main advantages of the inverter tube over the wafer tube occur as a result of the much higher energy (5 keV) of the electrons in the inverter tube compared to those in the wafer tube (300 eV).   This high energy enables a film to be deposited over the entrances to the channel plate which allows the passage of high-energy electrons but prevents the positive ions produced at the microchannel plate and phosphor screen from travelling back to the photo-cathode.   Thus this film enables the ion noise to be substantially reduced and also increases the life of the tube by preventing the ions from damaging the photocathode.

In the wafer tube then, noise caused by positive ion bombardment of the photocathode restricts gain to about 20 000 and therefore limits its usefulness at very low light levels.   This ion bombardment of the photocathode also re-duces the tube's operational life, especially when operated at screen brightness in excess of 1 Cd/m².

2.3. *Noise characteristics*

Another noise effect in microchannel-plate intensifiers arises chiefly because every primary photoelectron entering the plate does not produce an equal number of secondaries.   This variation results in a signal-to-noise ratio at the output of the channel plate which can be up to four times that of the input.   This is equivalent to reducing the illumination of the photocathode by a factor of four and is an especially critical factor in applications requiring the highest viewing sensitivity.   The visual effect of noise is that of image sparkling or scintillation under low levels of illumination.

The noise comparison between the channel plate and the cascade tube is not a straight factor of four.   Recent results [1] have shown that cascade tubes have noise factors as high as 1.6.   This is partly due to reflected primary electrons in the first stage and partly due to screen grain and fibre-optic non-uniformity broadening the pulse height distribution.

One way of minimizing the effect of the noise factor for channel plates is to place the phosphor screen close enough to the plate to sufficiently limit the spreading of the electron beam intersecting the viewing screen.   Another solution is to increase the persistence of the viewing screen.   This technique works well for fixed images but it has a tendency to blur the detail of rapidly changing images.

2.4. *Better image quality*

One of the best methods of measuring image quality is in terms of the modulation transfer function (MTF).

In the wafer tube, the distance between its photocathode and the surface of the microchannel plate has a significant effect on the tube's MTF.   Even under the most favourable conditions (with 0·1 mm gap and a voltage potential of about 300 V between the two surfaces), the spread of electrons leaving the photocathode limits the tube's image quality.

In the inverter tube, however, electrostatic focusing reduces this spread of electrons and MTF is mainly dependent on the number of channels per unit area, and on the proximity focusing between the channel plate and the phosphor screen.

Typical modulation transfer functions for each of the three types of image intensifier tubes are plotted in figure 3.   The cascade tube is Mullard model XX1060, and the inverter tube is Mullard Model XX1301.   Data for the wafer tube is taken from published data of other manufacturers.

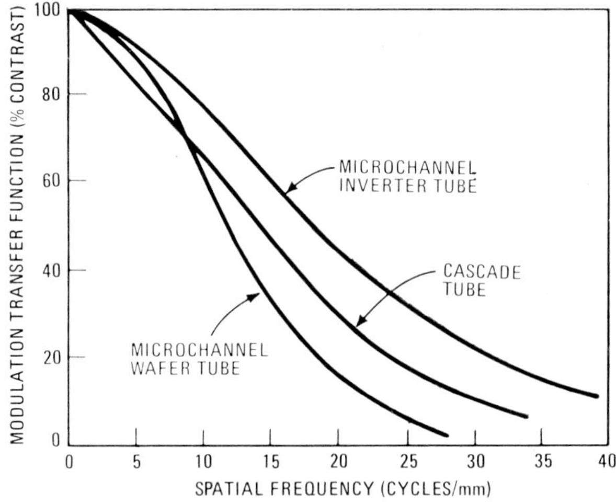

Figure 3.   Modulation transfer functions for three types of intensifier tube.

2.5. *Pin-cushion distortion*

Pin-cushion distortion arises from the practical need to minimize the length-to-diameter ratio of the intensifier tube.   The distortion occurs as a result of focusing the electrons on a channel plate which is flat rather than concave.   This distortion has been limited to 14 per cent for the XX1301 tube, which has a length-to-diameter ratio of 2 : 1.

The pin-cushion effect can be reduced, if necessary, by using an objective lens with compensating barrel-type distortion. When the inverter tube is used as an input to a television camera, normal television compensation circuitry can also reduce the effects of pin-cushion distortion.

2.6. *Highlight suppression*

One of the most useful properties of channel-plate intensifiers is the localized saturation of individual channels within the plate. Each channel, acting as an independent electron multiplier, can saturate without affecting neighbouring channels.

The effect of localized saturation can be used to advantage for highlight suppression in applications where there are bright point sources of light such as car headlights and street lights. In these situations, local channel saturation reduces the gain for bright sources without reducing gain in darker parts of the image. And, since recovery time from saturation of the channel is less than the response time of the eye, the highlight suppression is as effective on a moving image as on a static one.

2.7. *Applications*

Microchannel tubes were originally developed for military night-viewing binoculars. This system with the 18 mm tubes is shown in figure 4, where the

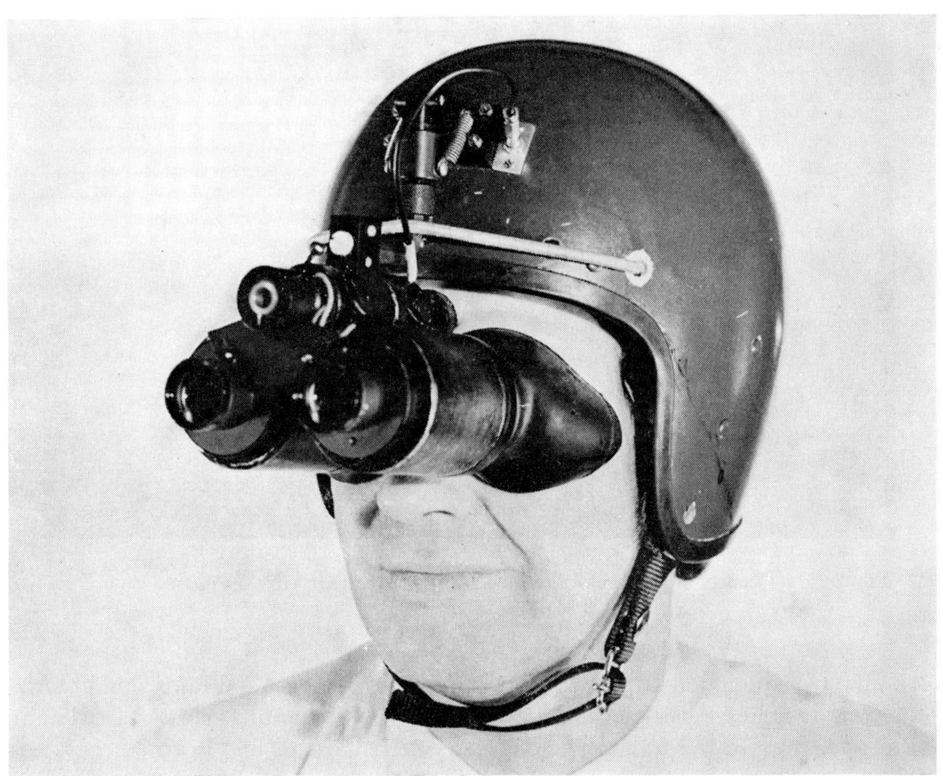

Figure 4. Night-viewing binoculars incorporating a microchannel tube as an image intensifier.

reduced weight and size advantage of the system can be seen. These tubes are also being used in a variety of other roles, ranging from hand-held starlight telescopes to pocket viewers suitable for general security work.

In addition to the 18 mm intensifiers, two other channel tubes are being made. One of these, the XX1330 series, has a 50 mm diameter cathode and a 40 mm screen. This type of tube has been developed for a vehicle driving sight, where the larger picture diameter gives an excellent image quality even at very low light levels. The other type of tube is the XX1360 series. This is a 25 mm channel tube with a very low distortion and has been developed as a light-weight military sight that can be hand-held.

The small size of the microchannel tubes has also encouraged the use of the tube as a preamplifier for conventional television cameras. Figure 5 shows the type of picture that can be obtained. The result was obtained from an

Figure 5.   Quality of picture obtainable from a vehicle driving sight used on a rainy night in autumn.

XX1330 tube operated at midnight in October in overcast and rainy conditions.

A new application is the use of intensifiers with scintillators as particle and beam detectors in the nuclear field. The intensifiers enable lower energy particles to be detected. One particular application, being developed at the Rutherford Laboratory, is the use of intensifier/scintillator combination for the real-time positioning of a narrow neutron beam.

3. Infra-red detectors

Today there is a wide range of proven infra-red detectors commercially available. Many of these types have been developed under military contracts and have been built to meet exacting performance and environmental requirements.

The detectors cover a wide wavelength range, extending from silicon detectors for the very near infra-red to TGS detectors for the far infra-red. Many of the detectors are photon detectors working with atmospheric windows and details of the performances of these are given in figure 6.

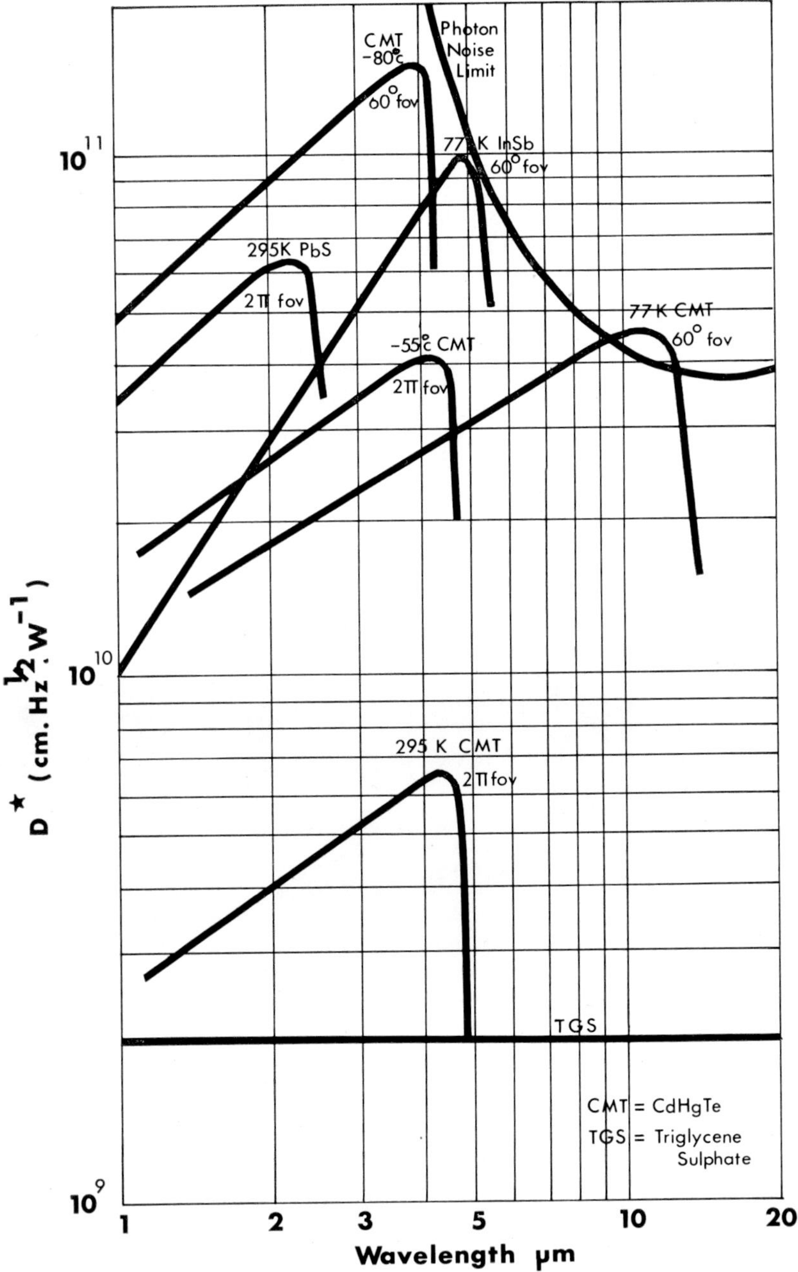

Figure 6. Spectral response curves of various infra-red detectors.

The latest range of photoconductive devices developed by Mullard are the cadmium mercury telluride detectors. By varying the ratio of cadmium telluride to mercury telluride, the energy gap and the wavelength of the detectors, can be varied. The material can be optimized for use both in the 3–5 μm window and in the 8–14 μm window. Background-limited performance can be obtained in the 8–14 μm window at 77 K, and at significantly higher temperatures in the 3–5 μm window.

In the design of any infra-red system the detector plays a central part. Its performance size and cooling requirements often affect the way in which the rest of the system is designed.

This influence has often meant that detectors have been manufactured to meet specific equipment requirements in order to achieve maximum system performance. This approach is expensive because little standardization is achieved and the trend is now towards providing the same detectors for a number of different applications. Standard types of 30 element 110 μm square on a 125 μm pitch array can already be provided for use with Joule-Thompson coolers and 192 element 50 μm square on a 62·5 μm pitch array are being developed.

In this paper, the flexible manufacturing techniques which enable detectors with widely different element configurations to be manufactured are described, together with some comparisons between the performance of 77 K 8–14 μm cadmium mercury telluride detectors, triglycine sulphate detectors and 3–5 μm thermoelectrically cooled cadmium mercury telluride detectors.

Finally a section has been included on the types of detector that have been developed for space applications.

3.1. *Detector arrays of 8–14 μm Cd Hg Te*

In order to achieve flexibility in the configuration of detector arrays, the detector element chips with contacts already deposited, are fabricated separately as shown in figure 7. The required array configuration is then detailed onto

Figure 7. Infra-red detector elements fabricated into an array.

the substrate and each element is then put in its correct place.   Photolithographic
techniques are then used to beam out leads from each of the detector elements
to the substrate pattern.   In this way arrays of up to 200 elements of 50 μm
size have been made.

For these arrays to be cooled to 77 K by Joule-Thompson or Leiden-Frost
systems, a dewar with multiple leads is needed.   To make the dewar of minimum
size, a relatively high density of leads is required.   The encapsulation construction chosen is the lead-in-glass form, in which the leads are taken up the
inner finger of the dewar in the thickness of the glass (figure 8).   A lead pitch

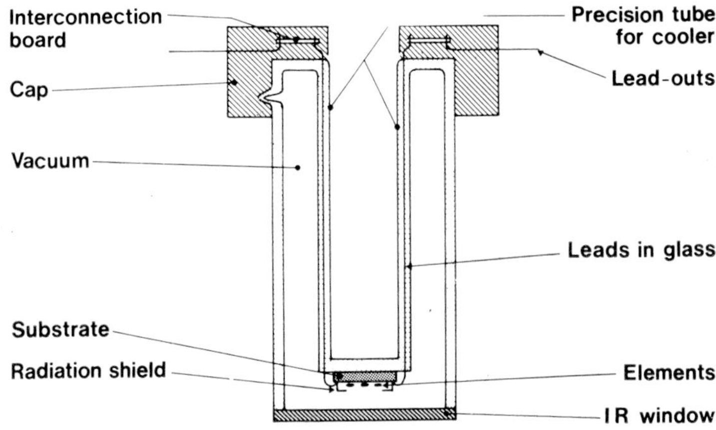

Figure 8.   Lead-in-glass connections to a dewar for detector arrays.

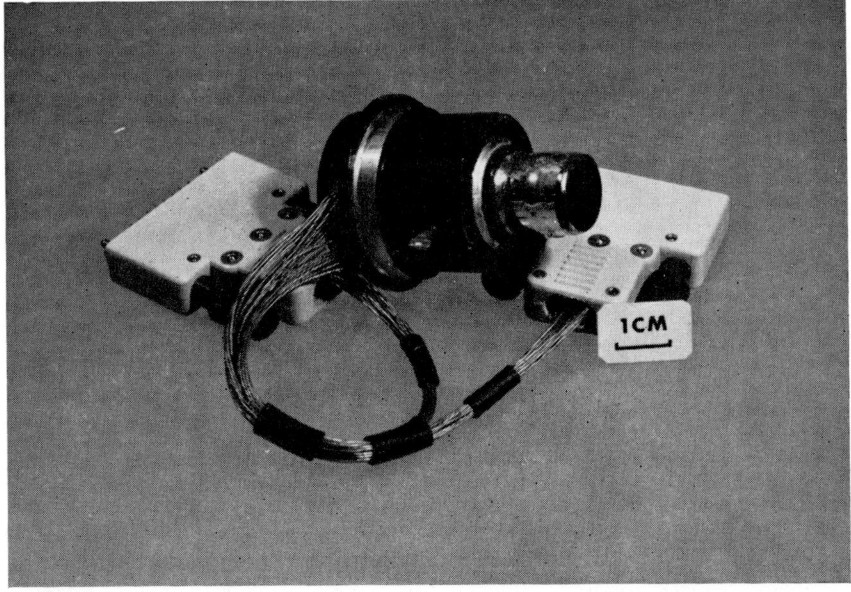

Figure 9.   Encapsulated 30-element detector array of CdHgTe.

of 400 μm is used. The leads provide bonding pads at the element end of the inner for connection to the sensitive elements, and at the other end may be connected to either flying leads, film leads or plugs.

Figure 9 shows a complete encapsulated 30-element array. The properties of such arrays are summarized in Table 1.

Table 1. Characteristics of the 30-element array shown in figure 9.

D^* (500 K, 5 kHz, 1)	: Min. $1 \cdot 10^{10}$ cm Hz$^{1/2}$ W^{-1}, typical $1 \cdot 6 \times 10^{10}$ cm Hz$^{1/2}$ W^{-1}
NEP (500 K, 5 kHz, 1)	: Typical $7 \cdot 5 \times 10^{-13}$ W
Noise (5 kHz)	: Typical 8×10^{-9} VHz$^{-1/2}$
Variation of responsivity	: Less than $\times 3$ range at constant bias
Resistance	: Typical 50 Ω
Bias current	: Typical 5 mA
Cut-off wavelength	: Greater than 11 μm

User experience has shown insignificant changes in element performance after many hundreds of hours of operation and cooling cycles.

Currently, arrays of 192 elements of 50 μm square on 62·5 μm pitch are being made. The encapsulation is naturally larger, although still using the lead-in-glass technique. The performance of an array of this type is indicated in figure 10,

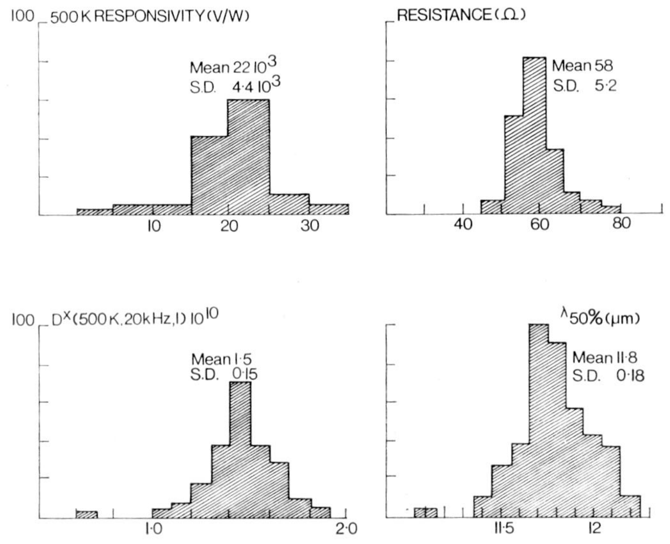

Figure 10. Performance of 192-element detector array of CgHgTe.

which shows the distribution of D^* responsivity, cut-off wavelength and resistance. The noise is similar to the 30-element arrays (typically 8×10^{-9} V Hz$^{-1/2}$).

Both these Cd Hg Te detectors are designed for operation with Joule-Thompson and Leiden-Frost cooling systems.

3.2. *Pyroelectric arrays using TGS*

Thermal detectors manufactured from the pyroelectric material triglycine sulphate (TGS) are very useful because they combine quite high performance with no cooling. Whilst these detectors can operate to very high frequencies, the maximum performance is obtained at low frequencies: 10 Hz to several hundred Hertz depending upon the element size. This means that for the parallel scan mode of operation where the upper bandwidth is substantially reduced, arrays of TGS can offer significantly improved system performance.

In order to obtain the best performance from TGS detectors it is highly desirable that each element should have the first-stage electronics built in: hence each device generally includes both elements and amplifiers.

Detectors of 128 elements have been constructed in a feasibility programme with an element pitch of $250\,\mu$m The elements are made of a single piece of alanine-doped TGS, with a common electrode along the length of the crystal on one side and interdigitated electrodes on the other side giving output connections on both sides of the array.

The element assembly is attached to the amplifier substrate which carries an FET and a load resistor for each channel and also provides the fan-out from the element pitch to a suitable pitch for connection to film leads or other outputs.

It is necessary to reduce capacitative coupling between adjacent channels to minimise cross-talk. An insulating earth plane is therefore used over the lead-out pattern. The capacitative loading of this earth lowers the responsivity but has a minor effect on the detectivity.

The performance of a 128-element array is shown in figure 11. The major part of the variation of responsivity across the array is due to variations in capacitance to the earth screen. Other non-systematic variations are less than 10 per

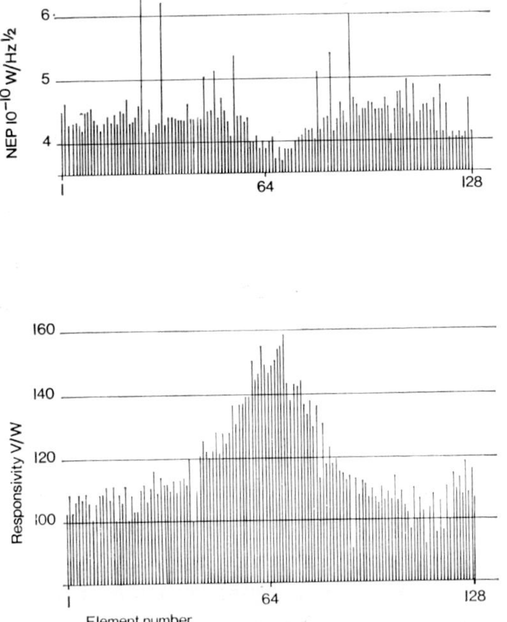

Figure 11. Performance of 128-element detector array of TGS.

cent.　The mean NEP (800 Hz) is $4 \cdot 2 \times 10^{-10}$ W.　At lower frequencies the performance improves, for example by a factor of 2 at 125 Hz or 4 at 12·5 Hz.　This performance is at present not as good as achieved in single detectors, where the typical NEP at 800 Hz is 2×10^{-10} W.

3.3. *3–5 μm CdHgTe detectors*

The performance of CdHgTe in the 3–5 μm band at temperatures from $-80\,°\mathrm{C}$ upwards has been described previously [2].　Recently, work has been carried out on detector elements for the 8–13 μm range at these temperatures and figure 12 gives information on performances that have been achieved.　Elements of this type can be made into arrays in a similar way to the 8–14 μm elements.

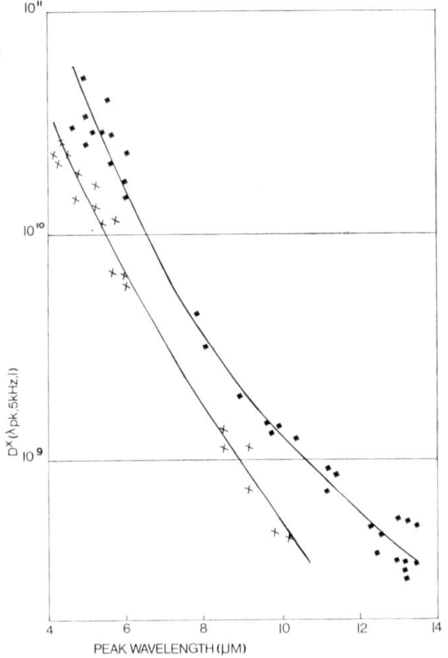

Figure 12.　Performance of CdHgTe detector in 3–5 μm band at 193 K (*top*) and 233 K.

Thermoelectric coolers provide a convenient form of cooling for these temperatures, and detector assemblies with small arrays of up to 7 elements have been made.　The cooling current required is 1·7 A, which will give element temperatures of $-60\,°\mathrm{C}$ from a $20\,°\mathrm{C}$ case temperature.　At lower temperatures still higher performance can be obtained.　At $-80\,°\mathrm{C}$ D^* 4 μm values of $1 \cdot 5 \times 10^{11}$ cm Hz$^{1/2}$ W^{-1} have been obtained.

3.4. *Comparison of detectors for thermal imaging*

A comparison of the different detectors described above may be made using the figure of merit M^* [3].　This figure indicates the temperature contrast in the scene and also makes allowance for atmospheric transmission.　A measure of this contrast is important for thermal imaging.

$$M^*(f_1, T_1, L) = D^*(\lambda p, f_1, 1) \int_0^\infty \mathscr{R}(\lambda)\, \tau(\lambda)\left(\frac{\partial W_\lambda}{\partial T\lambda}\right)_{T_1} d\lambda$$

where $D^*(\lambda p, f_1, 1)$ is the detector D^* measured at a frequency f_1 and normalized to a 1 Hz bandwidth (units : cm Hz$^{1/2}$ W^{-1}).

$\mathscr{R}(\lambda)$ is the relative response at the detector normalized to one at the peak. (Units: numeric).

$\tau(\lambda)$ is the atmospheric transmittance. (Units: numeric).

W_λ is the spectral radiant emittance (into a hemisphere). (Units: W cm^{-2} μm^{-1}).

T_1 is the black-body target temperature. (Units: degrees K).

L is the atmospheric range under standard conditions. (Units: km).

The performance in terms of M^* for a 2 km range of the detectors discussed is summarized in Table 2.

Table 2. Performance of detectors for thermal imaging.

Detector	Element temperature (K)	M^* (f, 295 K, 2 km) cm^{-1} Hz$^{1/2}$ K^{-1} (typical)	Frequency (Hz)
CdHgTe 30 elements	77	4×10^6	5000
CdHgTe 192 elements	77	3×10^6	5000
TGS 128 elements	295	1×10^4	800
TGS 128 elements	295	2×10^4	125
TGS Single element	295	2×10^4	800
CdHgTe 3–5 μm band	233	1×10^5	5000
CdHgTe 8–13 μm band	233	3×10^4	5000

3.5. *Detectors for space applications*

Infra-red detectors are becoming more widely used for space applications. These range from attitude-control horizon sensors to detectors specifically monitoring clouds of water vapour and ground temperatures.

TGS detectors have been developed for both horizon sensors and for applications such as the stratospheric sounding unit on Tiros N. TGS detectors have in the past exhibited excess noise phenomena such as Barkhausen noise. By changing the fabrication techniques these problems have been overcome and provided that the detectors are temperature-stabilized during operation, no Barkhausen or other excess noise is observed.

For space applications cadmium mercury telluride detectors are generally cooled to approximately 95 K using passive radiation coolers, at which temperature $D^*12\,\mu$m $> 2\cdot8 \times 10^{10}$ cm Hz$^{1/2}$ W^{-1} are obtained.

Figure 13 shows the type of picture that can be obtained with a cooled cadmium mercury telluride detector.   The picture was obtained from the VISSR satellite in a synchronous orbit.

Figure 13.   Picture obtained from a VISSR satellite using a CdHgTe array.

4.   Acknowledgments

Many of the devices discussed were developed for the Ministry of Defence Procurement Executive, DCVD.

Acknowledgment is made to members of the Development Departments EOD Mullard Mitcham and Southampton who have been involved in the activities described.

5.   References

[1] CHALMETON, 1971, *Acta Electronica*, **14,** 97.
[2] MORTEN, F. D., 1972, *Proc. European Electro-Optics,   Markets and Technology Conference*, Geneva.
[3] CHIARI, J. A , and JERVIS, M H., 1971, *Proc. Electro-Optics Conference*, Brighton.

Scanned arrays of solid-state detectors

P. C. NEWMAN

Allen Clark Research Centre, The Plessey Company Ltd,
Caswell, Towcester, Northants, UK

Abstract. This paper describes the mode of operation of single detectors and arrays in the near and far infra-red and concludes that the future of large-scale self-scanned imaging arrays for such applications as low-light television lies with charge-coupled imaging. In the far infra-red, single elements of PbTe and SnTe are being developed for use in, for example, laser anemometers.

1. Introduction

The other papers in this volume mostly discuss single detectors, channel plates (which provide area imaging without scanning), and television systems which use field scanning of electrons in a vacuum. I wish to direct my remarks particularly to arrays of solid-state detectors of radiation, in which the scanning can be provided by solid-state circuits. The band of wavelengths that these detectors cover can be divided into two parts: the far infra-red (say 3–35 μm), and the visible and near infra-red (say 0·4–1·1 μm). In the first case, the material of the detectors is not suitable for the fabrication of amplifiers, and hybrid arrays must be fabricated by techniques that are well known. In the second case, silicon is suitable for detection, amplification, and scanning, and fully integrated arrays can be produced. In all cases I shall describe individual detectors, before going on to discuss the manufacture and application of arrays.

2. Detectors for the far infra-red

2.1. *Triglycine sulphate*

Triglycine sulphate (TGS) is an organic, pyroelectric compound that can be grown from aqueous solution to form single crystals. These crystals are ferroelectric, and the incidence of radiation upon them causes a temperature difference with time. This temperature difference leads to a separation of charge between the two surfaces. With time, the rate of change of temperature and the separated charges will diminish, unless arrangements are made for the incoming radiation to be chopped, or for scanning to be used, with a period that does not exceed the thermal time constant of the device. Since they are essentially thermal detectors, the wavelength limits are set by the absorption coefficients of the TGS and of the window material. In the Plessey PSC222, using a KBr window, these limits are 2–35 μm. Using a 2×2 mm^2 element, the following values are given by the data sheet in the list of typical characteristics at 20°C:

$$D^*(500 \text{ K}, 10, 1) = 5 \times 10^8 \text{ cm Hz}^{1/2}\text{W}^{-1}$$

$$D^*(500 \text{ K}, 1000, 1) = 5 \times 10^7 \text{ cm Hz}^{1/2}\text{W}^{-1}$$

$$\text{NEP} (500 \text{ K}, 10, 1) = 4 \times 10^{-10} \text{ W}$$

(Peak emission from a 500 K black body is at 5·8 μm).

Performance figures similar to those for a single detector can also be obtained for linear arrays of 32 detectors, which can be fabricated quite simply (with element sizes down to 0·25 mm square). Each element requires an amplifier, which uses a JFET in a high impedance input stage; the operating frequency lies in the range 1–2000 Hz, and the amplifier is compensated to allow for the variation of the output of the detector with frequency.

Figure 1 shows the electrical circuit of a section of an array of TGS detectors, with amplifiers and scanning circuits. The latter use MOST multiplexing switches, made with silicon-on-sapphire technology, which are driven by standard TTL circuits.

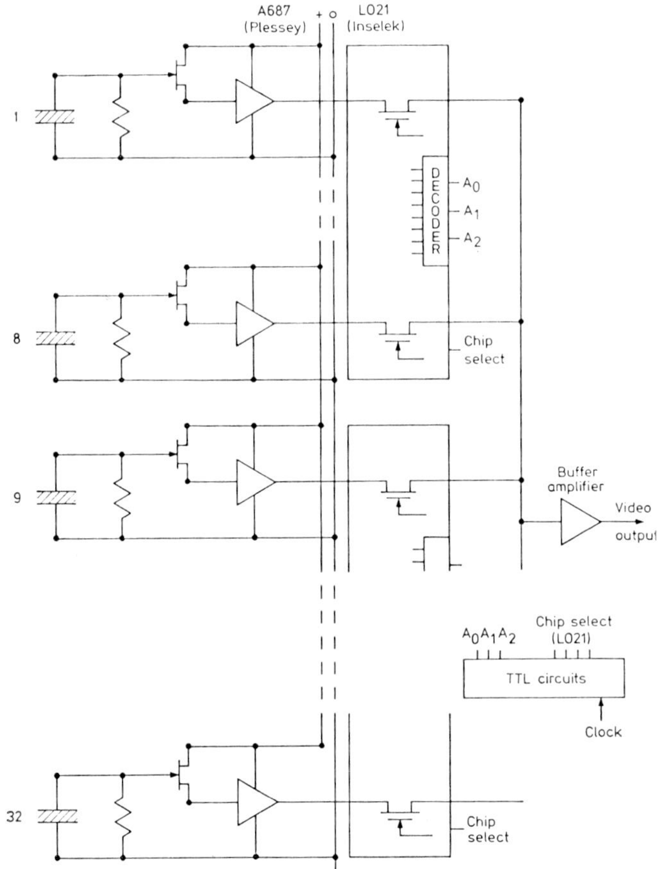

Figure 1. Circuit diagram for array of TGS detectors.

2.2. *Ceramic pyroelectric*

As well as the organic materials, typified by TGS, it is possible to make ceramic pryoelectric materials. These have a higher Curie temperature than TGS, but also require a higher temperature of preparation (for which hot-pressing is used). It is now quite feasible to make detectors out of polycrystalline material, which is given a highly polished finish. Directionality is induced in these layers by the process of ' poling ': this involves cooling the layer through the Curie temperature, under an applied electric field. After this treatment, the following typical performance is attainable at 20°C from a single 2×2 mm² element, behind a KBr window, as shown on the data sheet for the Plessey PPC522:

$$D^* \,(500 \text{ K}, 10, 1) = 1 \times 10^8 \text{ cm Hz}^{1/2}\text{W}^{-1}$$

$$D^* \,(500 \text{ K}, 1000, 1) = 1 \times 10^7 \text{ cm Hz}^{1/2}\text{W}^{-1}$$

$$\text{NEP} \,(500 \text{ K}, 10, 1) = 2 \times 10^{-9} \text{ W}$$

The Plessey detectors are made from compounds in the lead zirconate-titanate system. Other types of material that may be used for ceramic pyro-electric detectors include lithium tantalate and strontium-barium niobate.

Arrays of detectors of up to 128 elements are available; again, each element (which is now 0.25×0.20 mm² on a 0.25 mm pitch) is provided with an amplifier that has a JFET input stage. A typical range of frequencies of operation is 0.25–2000 Hz (this is set by the differentiating bipolar amplifier) and the spectral response covers the range 0.2–35.0 μm.

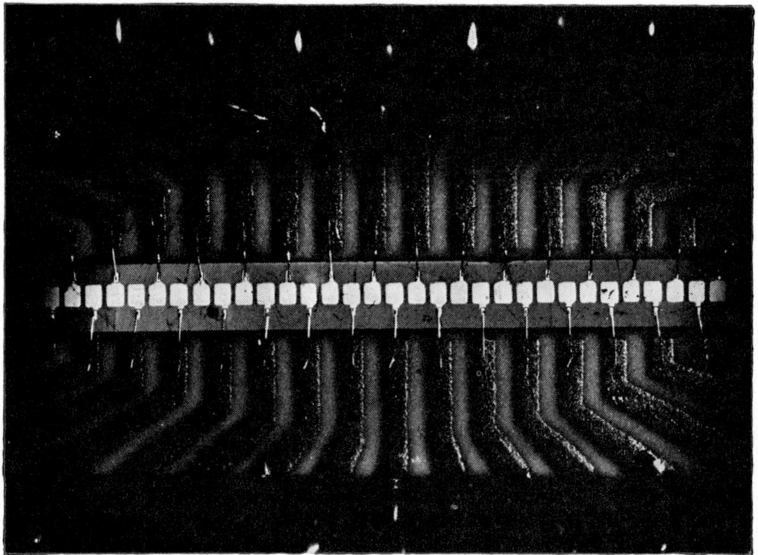

Figure 2.   Array of 32 ceramic detectors.

Figure 2 shows a photograph of an array of 32 detectors. At the present, the electrical circuit is identical to that used for the TGS detectors (see figure 1). However, it is expected that a new amplifier will be developed for both detectors, that will have a lower noise figure than the present one. By making the ceramic material sufficiently conductive, the bias resistor shown in figure 1 can be eliminated.

2.3. *Photovoltaic detectors*

The lead salts (PbS, PbSe, PbTe) and their alloys with tin salts (SnSe, SnTe) form a range of material for the fabrication of photovoltaic detectors in the far infra-red. Various combinations of these materials can cover the range 3–15 μm. Since they are detectors of photons (by band-to-band absorption), rather than of thermal energy, they require to be cooled for successful operation. The long-wavelength limit is set by quantum effects in alloys with a very small energy gap and the degree of cooling that is available.

Plessey have concentrated particularly on the fabrication of photodiodes in the lead-tin telluride system ($Pb_xSn_{1-x}Te$), using epitaxial material on PbTe substrates. These detectors are particularly good for covering the spectral range 8–14 μm, which includes one of the major atmospheric windows. This window is particularly important for the observation of scenes at around normal ambient temperatures, say 275–325 K, and for the detection of CO_2 lasers. The peak of emission from a black body at 300 K occurs at 9·7 μm, and CO_2 lasers emit at 10·6 μm.

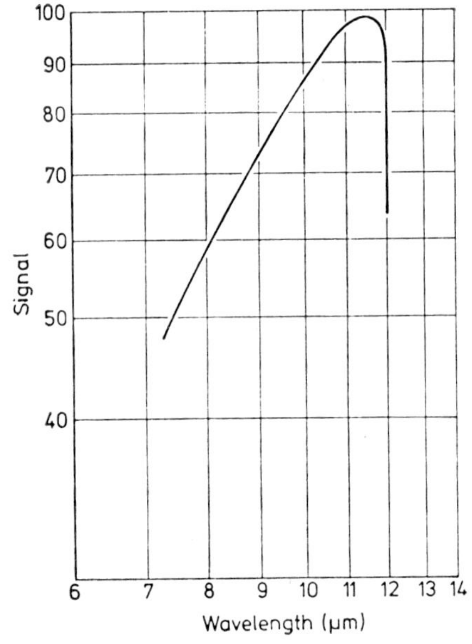

Figure 3. Spectral response of $(Pb_{77}Sn_{23})Te_{100}$.

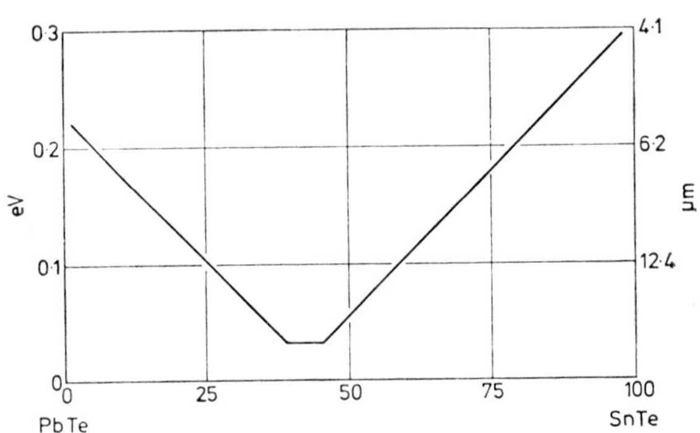

Figure 4. Variation of peak response with composition.

Figure 3 shows the spectral response of a detector made from $(Pb_{77}Sn_{23})Te_{100}$. The peak wavelength depends on composition in the manner shown in figure 4. The horizontal part of the curve indicates the longest peak wavelength that may be expected with cooling to 77 K.

Single devices have areas of $2 \cdot 5 \times 10^{-5}$ up to $2 \cdot 5 \times 10^{-3} \mathrm{cm}^2$. With the very low $1/f$ noise of (Pb, Sn)Te, detectivities (D^*) up to $6 \times 10^{10} \mathrm{~cm~Hz}^{1/2}\mathrm{W}^{-1}$ have been obtained in the detectors supplied commercially. The speed of response of these diodes is limited by the RC time constant of the input circuit to the following amplifiers; response times below 5ns have been achieved. A combined

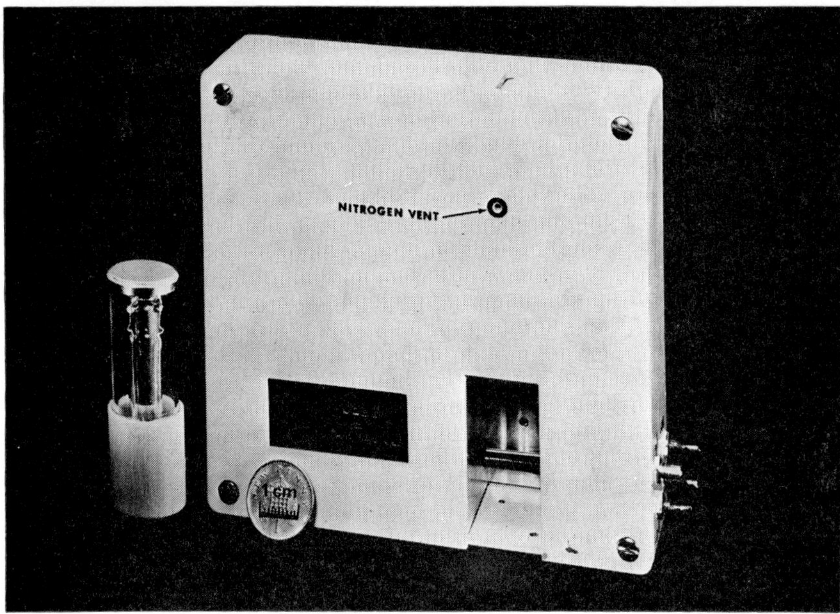

Figure 5. (Pb,Sn)Te detector and associated amplifier.

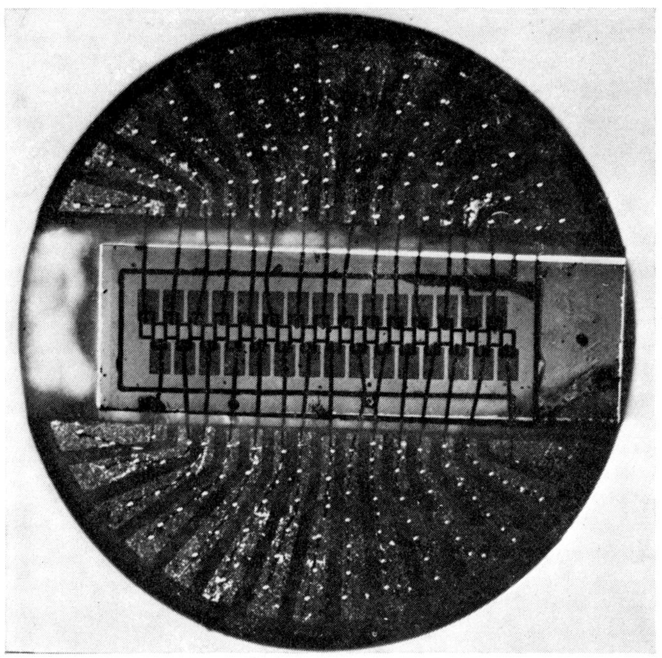

Figure 6. Array of 30 (Pb,Sn)Te detectors.

detector (with mini-dewar) and amplifier, as shown in figure 5, is now available with an operating bandwidth of 20 MHz and an effective D^* of 2×10^{10} cm Hz$^{1/2}$W^{-1}. The D^* of the detector itself is somewhat degraded by the noise of the amplifier.

Monolithic arrays of 30 elements are made, with element sizes from 50–250 μm square. These arrays are made by techniques of epitaxy, diffusion, photoengraving, etching and metallization that are derived from those used in integrated circuits, suitably modified for the change of material. A completed array is shown in figure 6. Very close matching of the spectral responses from all the elements is obtained, and the spread in current responsivity lies within ± 20 per cent.

A bipolar amplifier has been designed and made, which optimizes the matching of the detector at the input, and provides a reasonably low noise figure. The basic performance data of this amplifier are as follows:

Gain: 40–70 dB, preset by external components
Gain variation: ± 1dB from 0–70°C
Dynamic range: 50 dB for 3 dB compression
Upper cut-off frequency (3dB): > 250 kHz
Noise figure: 1·5 dB for common emitter input from 30–150 Ω source.

A circuit diagram of the amplifier is given in figure 7. Use of a JFET pre-amplifier stage could lower the noise figure still further, when suitable devices are freely available.

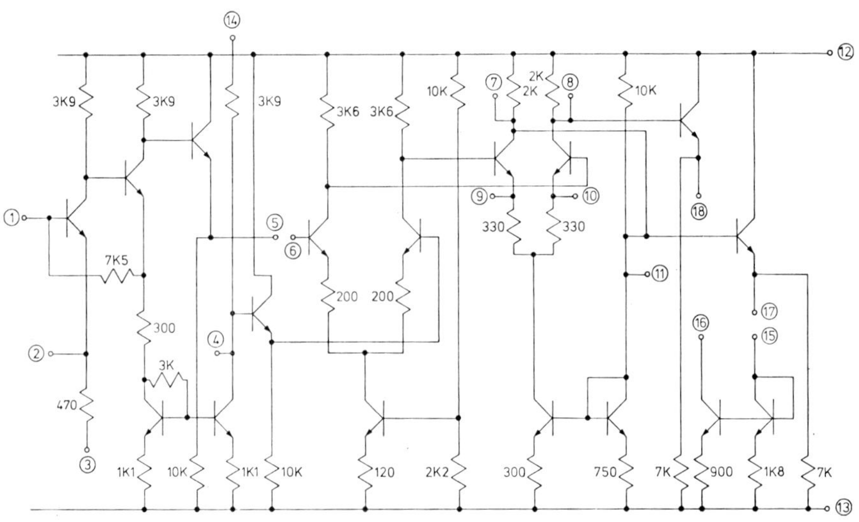

Figure 7.   Circuit diagram of amplifier for (Pb,Sn)Te detector.

3.   Self-scanned arrays for the visible and near infra-red

3.1. *Silicon arrays*

Silicon is one of the earliest semiconducting materials to have been fabricated into devices; its technology is by far the most advanced of all semiconductors, because of its use for integrated circuits. This use has developed, because it is possible to produce silicon dioxide layers of the required quality.

It is therefore extremely convenient that its spectral response as a photodetector covers the visible and near infra-red regions, with wavelengths from 0·4 to 1·1 μm.

As the techniques for making integrated circuits have improved, and the scale of integration has increased, so has the size of arrays of photodetectors in silicon. Many approaches have been tried, in obtaining solutions to the problem of integrating on one chip the detector, an associated amplifier, and a scanning circuit that will allow a parallel input of light to each element to be converted into a serial electrical output. Two of these have met with useful success, and these are described in the next two sections.

It is, of course, possible to think of many other semiconductors with energy gaps of the right value to give a spectral response centred in the visible region. None of them, however, lends itself as easily as silicon to the development of large-scale integration. Thus, all the arrays described in section 3 have the scanning circuits completely integrated with the detectors; this is in contrast to the arrays described in section 2, where scanning circuits, albeit solid-state, have to be made separately and formed into a hybrid assembly with the detectors.

3.2. *Arrays of photodiodes, scanned by MOSTs*

This was the earlier development, of the two that I am going to describe; work started at the Allen Clark Research Centre of the Plessey Company in 1966 [1]. The MOST (Metal-Oxide-Silicon-Transistor) can be very simply integrated, with a process that contains only one diffusion and four stages of photoengraving. The diode formed by the diffusion is found to be an excellent photodetector, with a typical spectral response shown in figure 8. This response curve arises very largely from band-to-band absorption in the silicon.

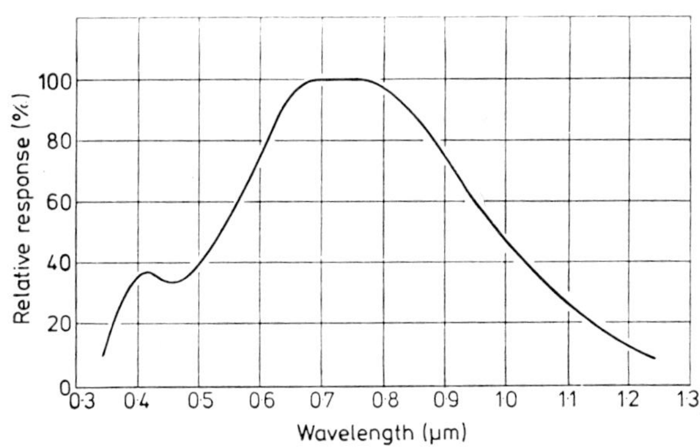

Figure 8. Spectral response of silicon photodiode.

At the blue end it is affected by the oxide layers above the diode; at the red end it is affected by absorption from impurity levels and by the diffusion length of minority carriers in the material. The response time is short, except possibly for the long-wavelength tail. It is possible to combine detectors, amplifiers and scanning circuits, without increasing the complexity of the process. Figure 9 shows a cross-section of a photodiode and a MOST. Whereas the MOST is isolated from the substrate by a reverse-biassed *p-n* junction, one side of the photodiode is automatically connected to the substrate, and this does put a certain limitation on the design of circuits. However, within this limitation, a number of improvements have been made over the course of years.

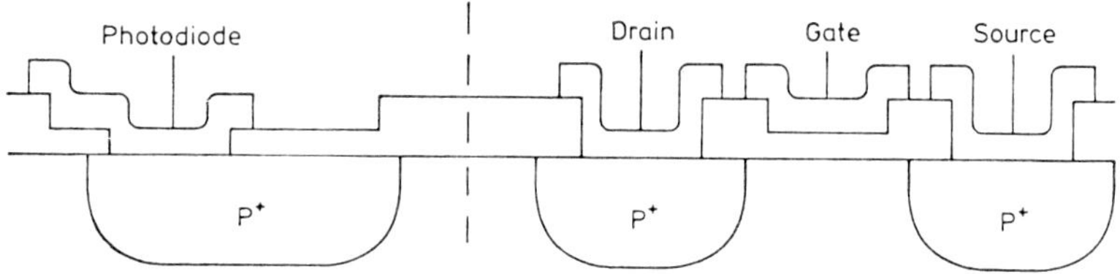

Figure 9. Cross-section of photodiode and MOST.

A basic decision must be made as to whether current sampling or voltage sampling is to be employed. The two designs for a single detector are shown in figure 10. In both cases, the photodiode is charged up to a given voltage,

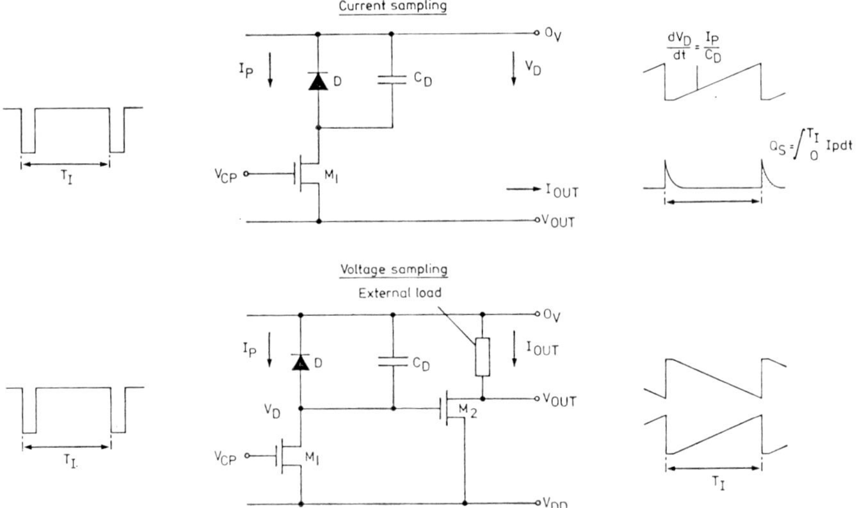

Figure 10. Current-sampling and voltage-sampling detectors.

and the incoming radiation discharges it. After a given period—the integration time—the state of discharge is measured, and the diode is recharged.

When detectors are integrated into area arrays, it is always necessary to use multiplexing switches at the end of each row. In the case of the voltage-sampling detector, an extra MOST must be added to each element (between M_2 and the load, which is now removed to the other side of the multiplexing switch) in order to carry out the voltage sampling just before recharging (current sampling occurs simultaneously with recharging).

Two particular improvements have been made in the voltage-sampling arrays. One is a double sampling technique, in which the video output is taken as the difference of two sample signals, before and after recharging. This removes the ' fixed pattern noise ' due to all the zero-level offsets in each element (except that due to variations in leakage current). The second is the use of a charge amplifier in each element, which consists of one extra MOST. This has the effect of reducing the leakage current of the photodiode (and hence a major source of random noise) to a value appropriate to the much smaller area of the extra MOST, whilst maintaining a gain of say, 10, for the voltage by which the diode is discharged by radiation.　Figure 11 shows the circuit diagrams

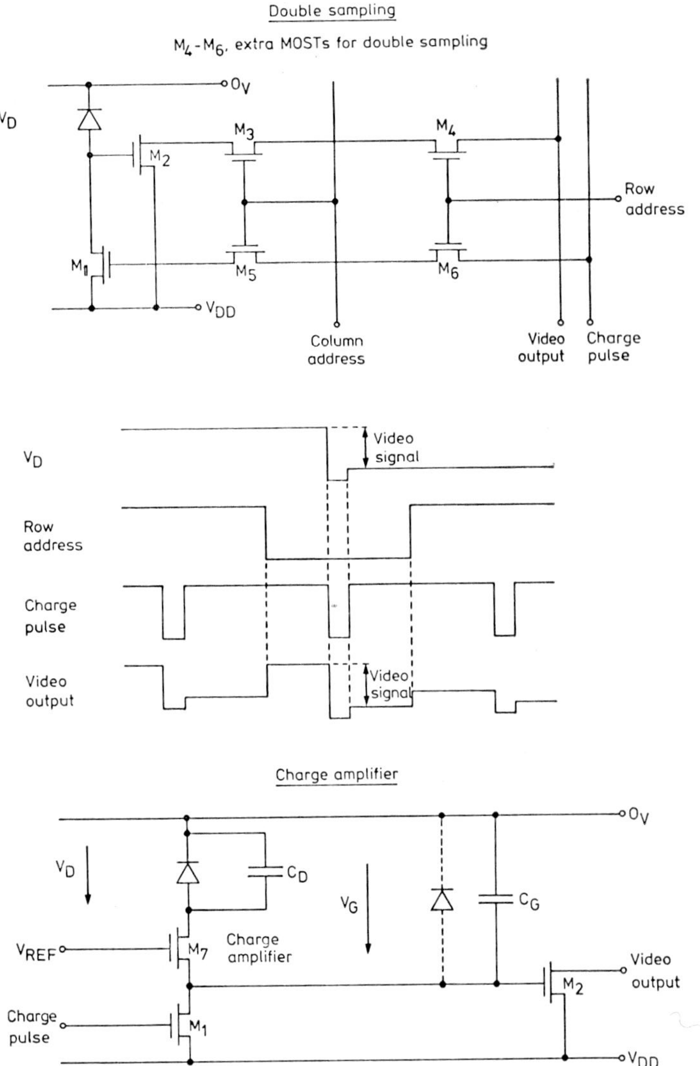

Figure 11.　Double sampling and charge amplifier for voltage-sampling detector.

of these two improvements, which can both be incorporated simultaneously in an array.　The extra MOSTs, compared with the simple voltage-sampling element, are indicated.

Some practical results that have been obtained are given in Table 1; in all cases the integration time for each photodiode in the array is taken as 0·1 ms. This will give varying rates of scanning, depending on the size and organization of the array.

Table 1. Characteristics of photodiode arrays

Manufacturer	Size of array	NEP* μW cm^{-2}	Responsivity μC J^{-1} cm^2	Dynamic range
IPL	64×64	133	0·2	200
Reticon	50×50	30	5·0	50
Plessey	10×10	75	0·7	100
Plessey	(Experimental)	$< 0·1$	250	≈ 1000

*NEP = noise equivalent power

3.3. *Charge-coupled imaging*

The concept of charge-coupled devices (CCD) was developed in the Bell Telephone Laboratories and announced in 1970 [2]. In these devices, a packet of charged minority carriers is held in a potential well in the semiconductor. By applying properly phased voltage pulses to a set of electrodes on the surface, the potential well may be caused to travel in a selected direction, dragging the packet of charge with it. This is illustrated in figure 12, for

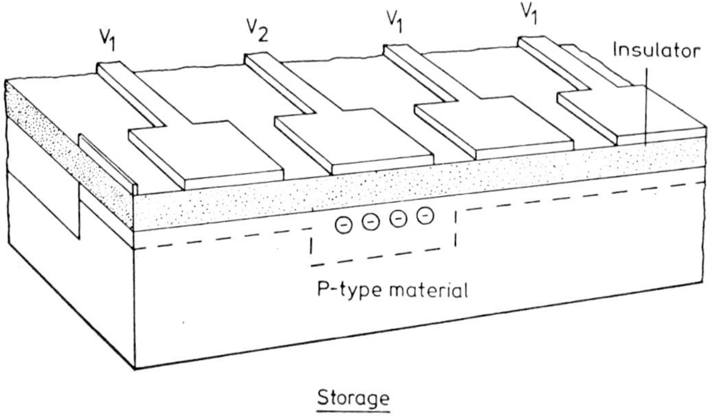

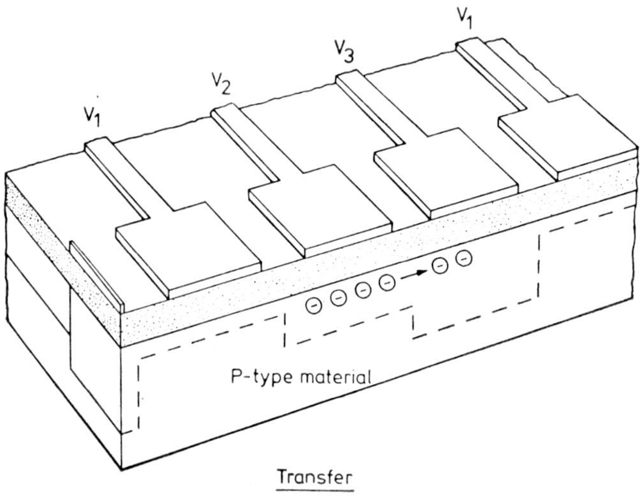

Figure 12. Three-phase charge-coupled device.

the case of three-phase operation; it is also possible to use four-phase operation or, by building the directionality into the electrode structure, only two phases. Without this structural signpost, application of only two phases of clock pulses does not define the direction of travel.

For the propagation and processing of electrical signals in CCD arrays, some forms of input and output devices are required. Most simply, these are *p-n* junctions prepared by diffusion into the silicon. More complex forms of output devices are also available, incorporating an amplifier on the chip. Developments since 1970 have also given the choice of having the depletion layer (which is formed by the surface electrodes) lying close to the silicon/silicon-dioxide interface (surface channel) or nearer to a large-area planar junction within the bulk of the silicon (buried channel).

Although CCD was originally designed for electrical inputs, it soon became obvious that the depletion layers formed by the electrodes could be used as photodetectors of an optical input. The minority carriers produced by this input can then be shifted in sequence to the output, through an on-chip amplifier if desired. There are several different approaches to this problem of shifting that I will describe later. This approach to imaging can be called charge-coupled imaging (CCI) [3].

There is a particular problem with CCI that does not arise with the arrays of photodiodes described in the previous section. To obtain charge transfer with very small values of inefficiency per transfer, say 10^{-4} to 10^{-5}, it is essential to make the gaps between the electrodes very small; if the electrodes are of aluminium (or other metal) then most of the incident light is prevented from reaching the silicon. There are three ways to overcome this problem:

1. Make the electrodes of semi-transparent silicon
2. Illuminate the silicon slice from the back
3. Provide photodiodes separate from the CCD shift register.

Each approach has its own advantages and disadvantages.

The type of shifting adopted by the Bell Telephone Laboratories is known as frame transfer, using three-phase clock pulses and metal gates, with a surface channel. This approach has also been adopted by RCA, who have produced a 512×320 array; this is the largest number of elements in any CCI device. The device consists basically of two halves. One half integrates the effect of the radiation falling on it, to give a pattern of stored charge. This pattern is then shifted rapidly into the second half, where it is further stored until required for read-out. Final read-out is by a linear shift register along one edge of the array. The light is detected through the gaps between the electrodes in the light-sensitive half. The organisation of the array is shown in figure 13.

A second type of shifting is called interline transfer and is used with separate photosensors, as shown in figure 14. This approach has been adopted by Fairchild, who have used it to make arrays up to 190×244. Two-phase clock pulses and polycrystalline silicon gates are used with buried channels; one of the gates covers the photosensor, which does affect the spectral response to a certain extent.

A third type of shifting is known as line transfer, but this has not yet been much used in CCI. In order to complete the discussion, figure 15 shows a hypothetical organization for a line-transfer array, using four-phase clocks

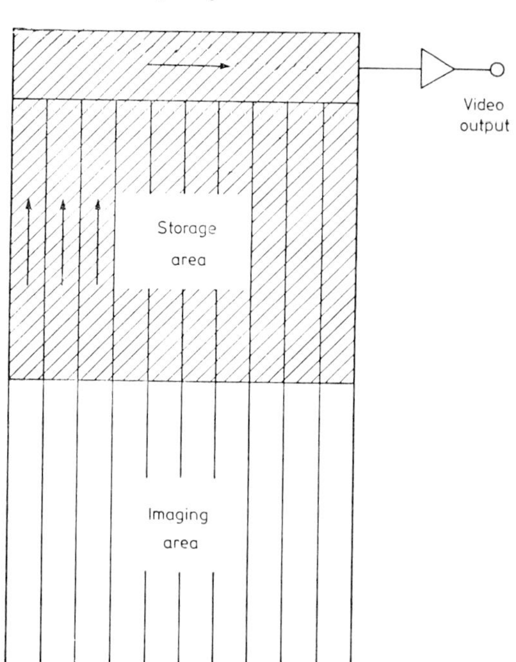

Figure 13.   Frame transfer CCI.

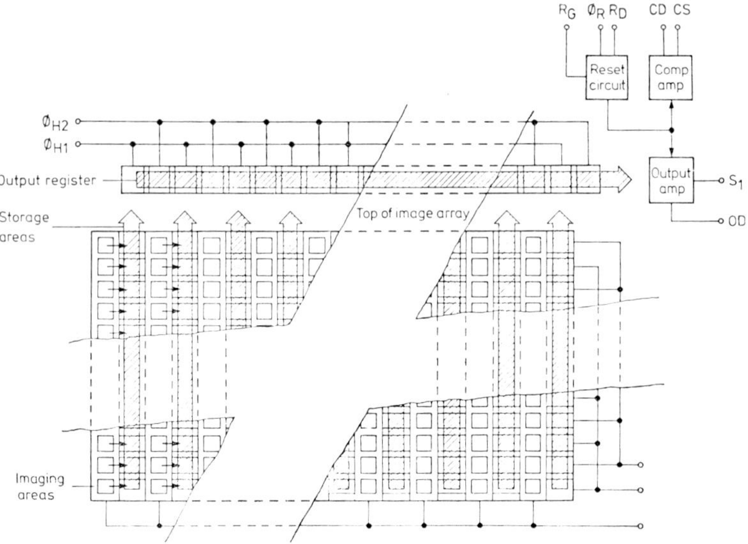

Figure 14.   Interline transfer CCI.

and metal electrodes, with back illumination. Illumination from the back would be essential, if very closely spaced electrodes are used. Plessey are currently working on a technique that will give spacings of less than $0.5\,\mu$m between long runs of metal.

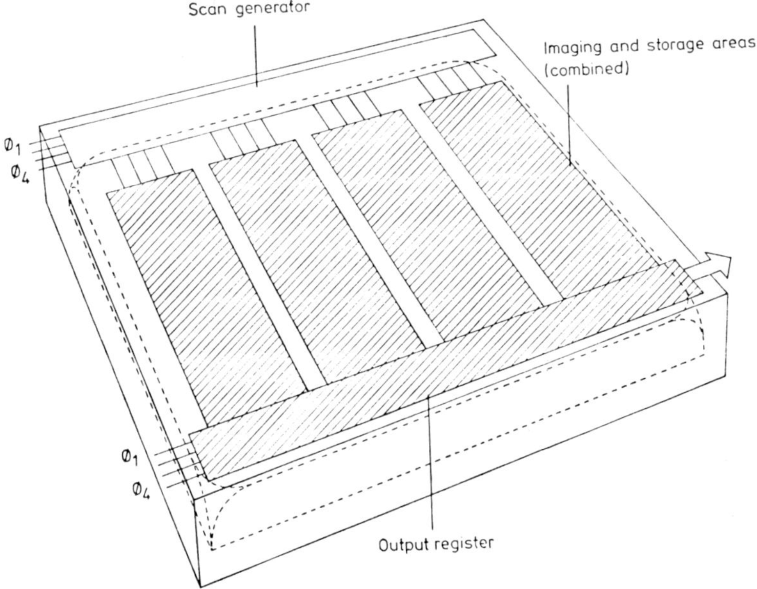

Figure 15.   Line transfer CCI (back illumination).

There is little doubt that the future of large-scale self-scanned imaging arrays in silicon lies with charge-coupled imaging, rather than with photodiodes and MOSTs.   One of the major reasons for this conclusion (which I have not discussed in this paper) is the greater performance in CCI; another reason is the considerably smaller area of silicon occupied by a single detector element in CCI.

4.   Conclusions

The type of imaging array described in the previous section is very suitable for low-light-level television.   Noise equivalent powers down to $10^{-3}\,\mu\mathrm{W}\,\mathrm{cm}^{-2}$ should be attainable for operation at room temperature.   Indeed a similar value is predicted for photodiode/MOST arrays, but it is more easily achieved with CCI.   Moreover, the operation of these solid state devices is very flexible, and they could very well be applied to such systems as laser-enhanced imaging, using range-gating etc.

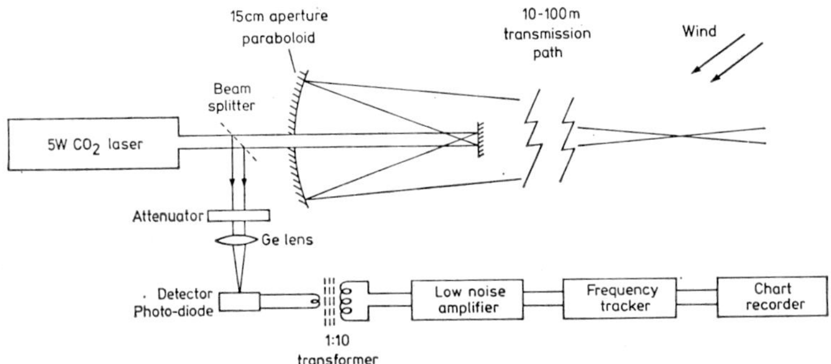

Figure 16.   Block diagram, $10\cdot6\,\mu$m laser anemometer.

As for the far infra-red detectors, single elements of (Pb, Sn)Te with their associated amplifiers are already being developed for use in laser anemometer and laser-ranging systems at $10\cdot6\,\mu$m. A block diagram of a laser anemometer developed at the Royal Radar Establishment, Malvern, is shown in figure 16[4].

5. Acknowledgments

Part of the work described in this paper has been carried out with the support of Procurement Executive, Ministry of Defence, sponsored by D.C.V.D.

References

[1] CAVE, K. J. S., WILLIAMS, B. D., and NOBLE, P. J. W., 1966, International Electron Devices Meeting, Late Paper, Washington. NOBLE, P. J. W., 1968, *IEEE Trans. Electron. Dev.*, **ED–15**, (4), 202.
[2] BOYLE, W. S., and SMITH, G. E., 1970, *Bell Syst. Tech. J.*, **49**, (4), 587.
[3] TOMPSETT, M. F., 1972, *J. Vac. Sci. Technol.*, **9**, (4), 1166.
[4] HUGHES, A. J., O'SHAUGNNESSY, J., PIKE, E.R., McPHERSON, A., SPAVINS, C., and CLIFTON, T. H., 1972, *Opto-electronics*, **4**, 379.

Review of UK activity in laser technology

G. M. CLARKE

Ferranti, Ltd, Scotland

Abstract. Lasers are already being applied in such diverse industrial systems as rangefinders, optical radar, cutting and welding as well as in research. This paper outlines the uses and suggests lines of future development.

1. Introduction

British companies have been active in laser technology from the early 1960s and the first commercially available European helium neon and ruby pulsed lasers were produced by Ferranti and by Barr and Stroud respectively. However probably more lasers were produced for experimental purposes in the 1960s than were used in systems. A considerable amount of time has elapsed before the applications in which coherent light has a clear advantage have emerged, and the necessary supporting technology has developed.

The applications can be divided into five major classes.

1. Optical radar
2. Energy concentration devices
3. Resonance applications
4. Optical instruments
5. Holography

Optical radar was originally of little interest to the military owing to the inability to penetrate weather but the situation has changed dramatically owing to the necessity for close-support operations with very expensive aircraft and ground vehicles.

Energy concentration devices for cutting and drilling difficult materials are based on carbon dioxide c.w. sources for bulk working and on pulsed ruby or neodymium-YAG for point applications such as resistor trimming. Both require the supply of considerable peripheral equipment in addition to the source.

The applications of lasers to chemistry is quite recent and has awaited the advent of continuously tunable sources. These are now available and contributing to the task of unravelling the structure of complex molecules in many fields.

The ability of laser sources to achieve very finely focused beams from small apertures leads to significant improvements in instruments such as surveying devices and their counterparts in metrology and alignment and is particularly important in scanning devices.

The principle of holography was established by Dr Gabor in the UK long before the invention of coherent sources which make it realisable. The drawback to its use so far has been largely the necessity for inclusion of photography with its attendant slow processing. Developments in electronic data handling and storage in recent years have brought nearer the removal of this barrier and real-time holographic systems for three-dimensional inspection and data processing are beginning to emerge.

2. Optical radar

Barr and Stroud have some 80 years' experience in distance measuring devices and were, in fact, the first firm in Europe to produce a working laser rangefinder.   The derivative to this equipment, known as LF2, is used currently to equip the Chieftain tank.   It offers the tank crew precise, instantaneous ranging and is a great improvement over the conventional, slow and somewhat less accurate optical rangefinders used on other Western tanks.

The LF2 head unit has been designed in a gunner's periscopic sight style in order to fit into the aperture vacated by the current sight, and consequently it traverses with the turret and elevates with the gun.   Basically, it consists of the laser rangefinder and a high powered telescope through which the gunner sights his target.   In order to range he simply places an aiming mark over the target and presses the Flash button.   Range is read off via a separate eyepiece which is situated so that the gunner's attention is never diverted from the target.   He then elevates the gun to an amount appropriate to the range.

The laser used in this system is a ruby type and has a power output of 1 MW. It can flash at a maximum continuous rate of 6 shots/minute.   A thermoelectric cooling arrangement maintains the crystal at room temperature over an ambient temperature range of $-26°C$ to $+70°C$.   The receiver normally measures the time to first return pulse but in the presence of smoke a last return pulse method is adopted which is completely satisfactory.

The environmental conditions inside a tank are severe and as a confidence check every laser sight unit produced must pass a final $40g$ shock test before leaving the factory.

A new system currently under development at Barr and Stroud is the hand held laser rangefinder.   It embodies all the basic features of currently available laser rangefinders yet is substantially smaller and lighter than most.   The equipment resembles a binocular in outward appearance and is completely self-contained.   The transmitter is a Pockel-cell Q-switched YAG laser which gives a power output of a few MW.   The receiver has a silicon avalanche photodiode detector and is fitted with a variable range gate to minimise short range clutter.   Range is given on an LED display.   Power is derived from integral rechargeable batteries which will each give several hundred shots.

The hand held laser rangefinder will range out to more than 5 km under average conditions and it weighs approximately 2·5 kg.   For small long range targets out to the maximum range of 10 km it is preferable to use a tripod for precise aiming.

Ferranti in Scotland, initially established in 1943 for the manufacture of gyro gunsights, now covers the full range of electronics with particular emphasis on equipment for civil and military aviation requirements.   The Electronics Systems Department, based in Edinburgh, has 20 years of experience in the design, development and production of ground and airborne equipments and is equipped with full flight and environmental test facilities.

The Laser Systems Group is an active part of the Electronic Systems Department and in addition to the Laser Ranger and Marked Target Seeker, is engaged on other military projects including the Laser Target Marker/Ranger for the British Army and the Laser Ranger/Marked Target Receiver for the Panavia MRCA.

Ground defences against air attack have become more and more effective in recent years. In Europe and a growing number of other theatres the strike pilot is now faced with advanced and often unpredictable anti-aircraft technology in areas where he is also liable to be denied both physical and ECM air superiority. At the same time, both man and machine have become extremely costly, since they are both expected to achieve more than ever before in an increasingly difficult environment. The attrition rate liable to be met in combat using conventional pull up and dive attack techniques becomes increasingly high.

The only alternative, given the above environmental conditions, is for the pilot to fly low and fast. Unfortunately, although this philosophy undoubtedly gives maximum protection from the greater part of most air defence systems because of the element of surprise and ground shielding, it also gives the pilot and weapon aiming system an extremely difficult task, since the target will usually only be visible for a very short period late in the approach. In close air support missions the target is also likely to be fairly small and well camouflaged. Indeed, without direction from a Forward Air Controller (FAC) the pilot will probably never sight the target at all and his attack will be abortive.

There are many benefits other than just accuracy that result from the low beam divergence characteristic provided by lasers. A laser range-finder measures distance almost instantaneously, thereby reducing the pilot's target tracking time so that he can afford to sight his target much later in his approach. Of course, this point is valid for radar, but the latter does not have the same inherent accuracy. Low beam divergence also permits accurate range measurement at the very shallow sight-line grazing angles that are associated with low level weapon delivery, and this is particularly difficult to achieve with radar. Being an optical device, the laser is more easily and more accurately harmonized than radar, especially to additional sensors like a marked target seeker. Finally, a laser is much more compact than an equivalent radar system. At the same time it is true to say that the current generation of laser systems will not operate in very poor visibility (cloud cover etc, not necessarily darkness) but in these conditions low level aircraft manoeuvring becomes hazardous in any case.

The Marked Target Seeker also deserves special mention, since it greatly assists the strike pilot in his difficult task of locating the correct target at first pass whilst still remaining low and fast. When air support is required, an aircraft is assigned to a battle area and is available to the FAC who designates a particular target with his laser. During the approach, the aircraft seeker detects scattered energy from the target and tracks the source of maximum energy. The pilot's head-up display is deflected in order to lie over the target in elevation and azimuth and he is then required simply to follow his display to the specific target. The Ferranti equipment combines marked-target seeking with laser ranging, with the result that it now becomes possible for the pilot to carry out his attack without ever actually sighting his target. At the same time his workload is reduced.

The extra complexity involved in adding the Marked Target Seeker to the basic laser ranging system is something of the order of only 20% and the penalty is far outweighed by the advantages to the strike pilot and ultimately the man on the ground.

3. Energy concentration devices

Two companies have divisions applying carbon dioxide lasers to industrial cutting.   Work on industrial laser systems started at BOC in their research and development laboratories in 1968.   In 1970, the first commercial and production laser system was sold for cutting out dieboards for the cardboard-carton industry. BOC Industrial Laser Systems expanded in 1972 to cover gas and solid-state laser systems based at the BOC Laser Centre at Daventry, Northamptonshire. Recently, the laser group has combined with the BOC Electron Beam Department under the new name of BOC Limited, Industrial Power Beams.   They can now offer a complete range of laser and electron beam systems for materials working.

Ferranti Ltd. in Dundee originally produced large CO_2 lasers (exceeding 1 kW) for cutting purposes but foresaw the need for more easily steerable sources to avoid the necessity for moving large workpieces.   The result is the Ferranti MF400 which folds the optical path twelve times to obtain a unit only 1·5 m long producing 400 W c.w. and more than twenty systems have been delivered to a wide range of customers.   A typical application is that of Plascut of Rotheram who use the laser head in conjunction with numerical control from a Ferranti Freescan digitiser which enables the machine to follow any line.   The UK Atomic Energy authority at Culham has done considerable work on the steering of high power lasers and provides a consultancy service.

Ferranti Ltd., is active in producing diamond drillers based on 1J pulses. The diamond die driller type DDD.1 has been specially designed to the requirements of the wire drawing industry.   The system employs a solid-state laser whose output is focused onto the die.   Closed-circuit television enables the operator to position the die exactly beneath the focus of the laser and also to observe the drilling as it progresses.   The diamond is mounted in a metal collar and held in a chuck which is rotated to obtain the best hole profile.   By off-setting the axis of rotation of the die with respect to the laser beam, larger diameter holes can be trepanned.

A similar application of pulsed lasers is available from International Research and Development at Newcastle.   This is the laser microwelder.   This hand-held laser can produce miniature welds in locations inaccessible to conventional welding machines and can often outperform them by welding dissimilar and thermally incompatible workpieces.   For example, it has been used to weld 0·1 mm diameter Kovar voltage probe wires to the relatively massive silicon iron stator core of a simulated 2000 MW generator.   Here the ability to take the laser to the workpiece was essential.

The compactness and portability of the system make it ideal for such on-site uses as the attachment of electrical contacts, thermocouples and strain gauges, for example in aircraft construction and testing and in the design and testing of heat exchangers.

4.   Lasers as research tools

Two powerful instruments for molecular analysis are available from Precision Devices, Malvern, England.   The laser photon counting molecular analyser, type K5600 uses light scattering spectroscopy with photon counting detection and digital storage.   The Malvern System 4300 is a photon correlation spectrometer for study of molecular hydrodynamic motion, molecular weight, rotation and shape.   The company also produce other components which have been developed from the activity of the Advanced Laser Group of the Royal Radar Establishment, Malvern.

Organic flash pumped tunable dye lasers for research are produced by Electro-Photonics of Belfast and are the subject of a paper in this volume. Edinburgh Instruments, which is closely associated with the Heriot Watt University, produces highly stable carbon monoxide and carbon dioxide lasers. The carbon monoxide tunable laser is the key to exploitation of the ' flip spin ' spectroscopic technique.

5. Laser based optical instruments

Scientifica is now the major UK supplier of helium neon gas lasers up to 40 mW. It includes a low-power educational laser and a compact portable alignment laser with and without integral power supply. They also market a standard holography system for non-destructive testing, stress and vibration analysis etc. A new system has been produced using laser beams to align x-ray machines on hospital patients.

Laser anemometer 6200 by Precision Devices uses digital photon correlation techniques to make greatly improved measurements of flow velocities and turbulence and overcomes many limitations of previous laser anemometers.

A problem important in the gas industry is the detection of small quantities of methane. The $3 \cdot 39 \, \mu$m line of HeNe lasers coincides with an absorption band of methane and an instrument based on this has been produced by International Research & Development (IRD) in conjunction with the Safety in Mines Research Establishment.

IRD have also been active in the development of Videoscan in conjunction with Loughborough University. This is a speckle pattern holographic system for non-destructive testing or vibration analysis.

6. Conclusion

The above brief survey indicates that proven laser applications are now numerous. They range from simple measuring techniques with new-found accuracy to complex research activities. Although the appearance of new frequency lasers is now less likely there is little doubt that new applications will continue to appear.

7. Acknowledgements

The author is indebted to information from the following suppliers.

Barr & Stroud, Caxton Street, Anniesland, Glasgow, G13 1HZ, Scotland.

B.O.C. Ltd., 10 Middle March, Long March Industrial Estate, Daventry NN11 4PQ.

Edinburgh Instruments Ltd., Research Park, Riccarton, Currie, Midlothian EH14 4AP, Scotland.

Electro-Photonics, The Cutts, Dunmurry, Belfast BT17 9HN, Northern Ireland.

Ferranti Ltd., Electronic Systems Dept., Ferry Road, Edinburgh EH5 2XS, Scotland.

Ferranti Ltd., Professional Components Dept., Laser (Sales), Dunsinane Avenue, Dundee DD2 3PN, Scotland.

International Research & Development Co., Fossway, Newcastle-upon-Tyne, England NE6 2YD.

Precision Devices & Systems (UK) Ltd., Spring Lane Trading Estate, Malvern, Worcestershire WR14 1AL, England.

Scientifica & Cook Electronics, 78 Bollo Bridge Road, Acton, London W3 8AU, England.

The application of the minicomputer to high speed on-line optical inspection systems

G. M. CLARKE and J. BEDFORD
Ferranti Ltd., Thornybank, Dalkeith, Scotland.

Abstract. A laser scanning analyser used in conjunction with a minicomputer can scan a wide variety of materials moving at high speeds and can detect flaws down to the size of the material's surface roughness. Links with larger general-purpose computers are now being investigated which would increase the power of the system still further.

1. Introduction

The Model 71 B laser scanning analyser, one element of which is shown in figure 1, is a high resolution optical scanning system initially designed for on-line

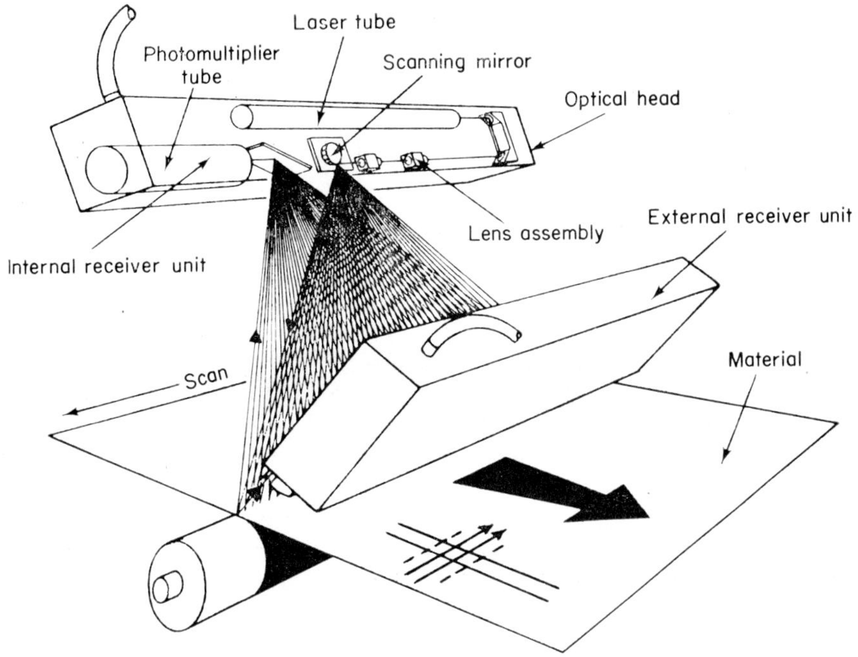

Figure 1. Laser scanning analyser, Model 71 B, was designed originally for on-line inspection of strip moving at high speed.

optical inspection of materials manufactured in strip form and moving at high speeds. It has facilities for size and contrast classification and for profiling round isolated flat objects on conveyors. The optical modes cover backscattered light, specular reflection and transmission, with a total of twelve receiver inputs. In maximum-width form these are stacked into six pairs covering 4·2 m width, at full resolution of the order of 0·1 mm.

When the shape of objects or defects becomes apparent, more powerful data processing facilities are required.   One of the major problems in providing this in a cost-effective manner is the high data rate of the system, which exceeds the direct store input of present-day computers.

The paper describes the capabilities of a laser scanning analyser using a general-purpose minicomputer.   It covers the design of the special interface necessary, techniques to extend the present receiver performance for use with a digital computer, and a description of the more relevant program necessary to drive the total system.   The elements of the system are shown in figure 2.

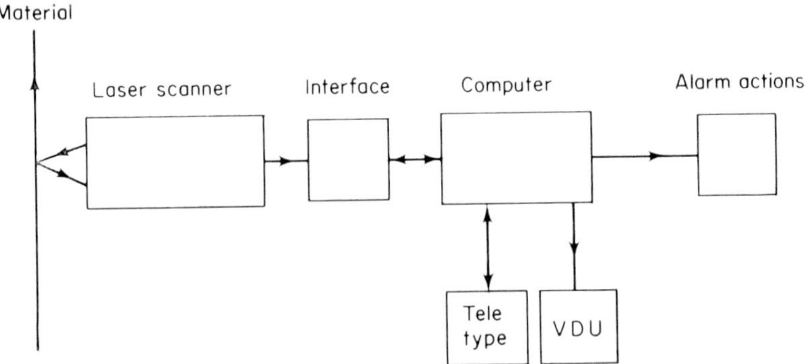

Figure 2.   The laser scanning analyser can be used in conjunction with a minicomputer, which greatly improves its recognition capability.

In order to exploit the full potential of the digital computer for on-line analysis of information collected by a laser scanning analyser, the interface described in Section 2 is used to split the inspected material into an orthogonal matrix of rectangles of equal area, oriented to the across and down-web directions. The size of these rectangles or zones is chosen to provide the required level of positional discrimination of a fault, size discrimination of a fault, or isolation discrimination between faults, whichever requirement is the more severe.

This preset level of discrimination is adjusted to suit the application, since it is desirable for the computer to be freed from manipulating data of unnecessary detail.

Zone characteristics are converted from scanner information into computer data, and this allows the computer to visualize the inspected sheet as a zone-matrix image containing analyser signal-intensity information.   Section 4 on computer programs discusses ways in which this information can be used to check inspected material.

The features of interest to the computer, having a larger range of character-istics than could be handled with a stand-alone analyser, require the improvement of the optical receiver characteristics using novel electronic techniques, as described in Section 3.   This allows the detection of low-contrast images of the same order of size as the receiver width.

2. The computer interface

The maximum data capture of a high resolution scanning analyser with a bandwidth of 20 MHz, observing a highly structured object, is about 100 times faster than the rate at which information can be assimilated by a general-purpose computer. The incoming data is therefore reduced to allow the computer to use it in real time. This is done in various ways:

(*a*) The computer analyser is normally used to observe material with few or no permanent features apart from edges. The information arriving at the computer therefore is purely related to fault conditions in the inspected material. If the computer analyser is inspecting components, there is a limit to the fineness of detail fed to the computer related to the rate of presentation of components.

(*b*) Inspected material zones may, and usually do, contain more than one scan (see figure 1). The interface accumulates the characteristics of each of the zones from a number of scans, transmitting these to the computer when a row of zones has been completely traversed.

(*c*) The ultimate discrimination level provided by the interface determined by the size of zones, being lower than the ultimate analyser resolution, provides a degree of data-rate reduction.

(*d*) If the texture is captured, the interface does not pass count information to the computer unless the count value per zone exceeds a predetermined limit fixed in the interface.

The Ferranti interface has discrimination levels from 256 to 1536 zones across the web, predetermined by wire links. The down-web discrimination ranges from 1 to an infinite number of scans. For multiple scan head systems the interface can take information from up to six such asynchronous scan receivers.

Two forms of zone data may be collected:

(*a*) A binary valued overthreshold word describing whether the analyser signal has been greater than a particular preset threshold level during the time the laser spot has been incident upon any part of the zone.

(*b*) For the inspection of materials with a noisy surface texture where the overthreshold word would be set for every zone, texture-count words are used. The count word is incremented each time the analyser signal crosses a predetermined threshold in a pre-determined direction at any place within the zone.

Recirculating digital binary stores are used to accumulate input data over a number of scans and upon an external command this is read in real time into a buffer store. The buffer store records zone across-web locations along with data for zones containing information of interest. After the buffer store loading is complete, the computer can gain access to it, and reads out information at a rate determined by the software activity. The input recirculating stores are continuously operating so that no data is lost. While unloading data into the buffer, the recirculating stores are simultaneously reset to allow the collection of fresh information from the analyser.

3.   Optical receiver compensation

The wide-angle optical receiver used with the laser scanning analyser has a non-constant gain which is a function principally of the position and direction of the light coming from the inspected material. As the laser beam sweeps across the material, so the receiver response changes.   Modern receiver design reduces much of this gain variation for specific inspection tasks using variable mechanical geometry, but for a universal fixed receiver, a gain variation of up to 1 : 2 can still occur over the active receiver length when different materials are scanned.   This must be electronically compensated.

The receiver response from a plain featureless sheet of inspected material is directly proportional to the receiver gain-function.   This signal must be eliminated since it is not related to the features which may appear on the background.   Furthermore, any feature signals must have their amplitude modulated electronically to compensate for the changes in the optical-receiver sensitivity. One of two techniques may be used to carry out this task.

3.1.   *Where features are small in across-web dimension* compared with the receiver width, the feature signals may be isolated from the featureless receiver signal because of the different frequency bands.   This technique is satisfactory for features smaller than approximately 1/20th of the optical-receiver width. Electronic gain compensation is provided by an amplitude modulator controlled by the featureless receiver signal filtered from the total response: the feature band signal, also filtered from the total response, passes through the modulator (figure 3).

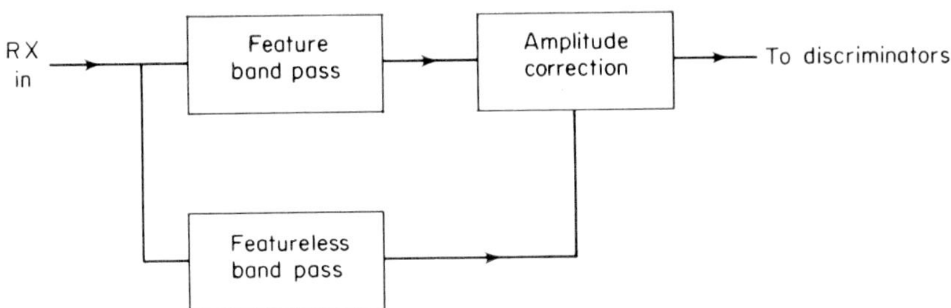

Figure 3.   Features that are small in the across-web direction can be detected by isolating the feature signals from the featureless background.

This arrangement is very effective when the features of interest are small and have sharp edges.   Figure 4 (*a*) shows an uncompensated receiver in action on a material which differs from the surface with which it was optimized.   The scene consists of a sequence of black stripes of increasing width repeated across the scan.   Figure 4 (*b*) shows the equalization effected by the circuitry of Figure 3, which is completely effective in restoring the balance across the scan. However it can be seen that there are white overshoots present owing to the missing low-frequency components.

While there are circuit methods of preventing these white overshoots becoming confused with true white signals, this is only one manifestation of the difficulty.   More serious is the case where the size or lack of a sharp edge to the feature gives frequency overlap between the feature band and the receiver-response band.   This circuit then either fails to detect such features or responds only at the edges, and in the latter case gives leading and trailing edges with opposite polarity.   The effect of this on a solid feature is as shown in figure 5.

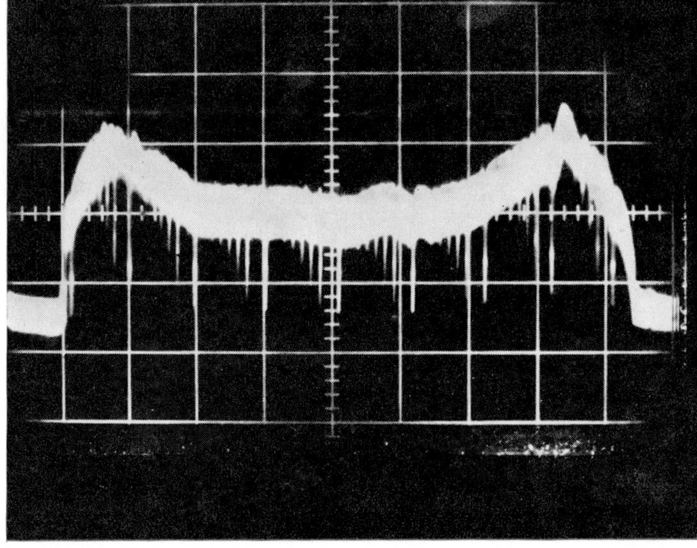

(*a*)

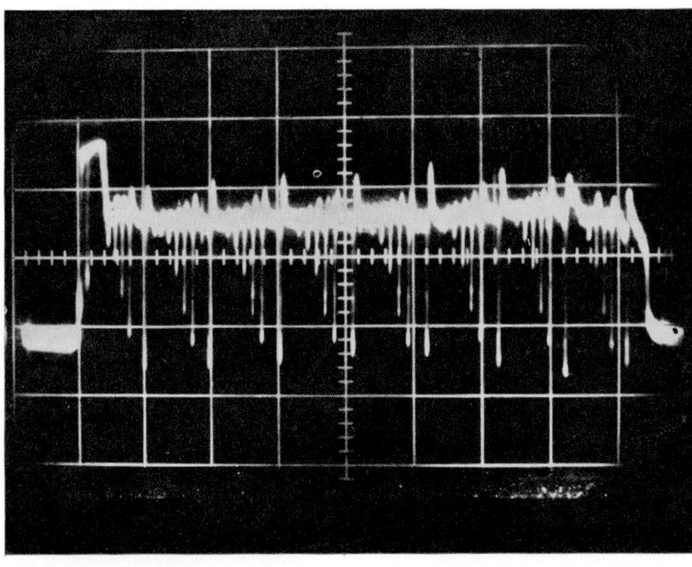

(*b*)

Figure 4.   Signals from small features with sharp edges.   (*a*) Uncompensated receiver looking at an unfamiliar surface.   (*b*) The improvement due to the circuits of figure 3.

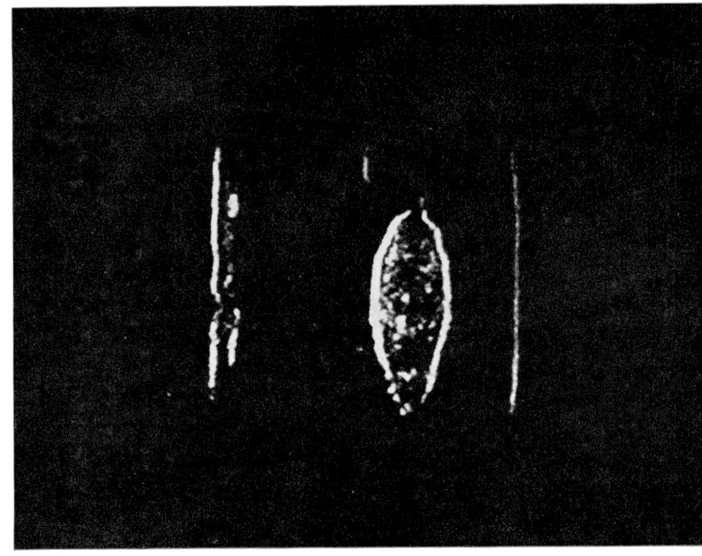

Figure 5.   Signals from a large feature where the feature frequency-band overlaps the receiver-response band.

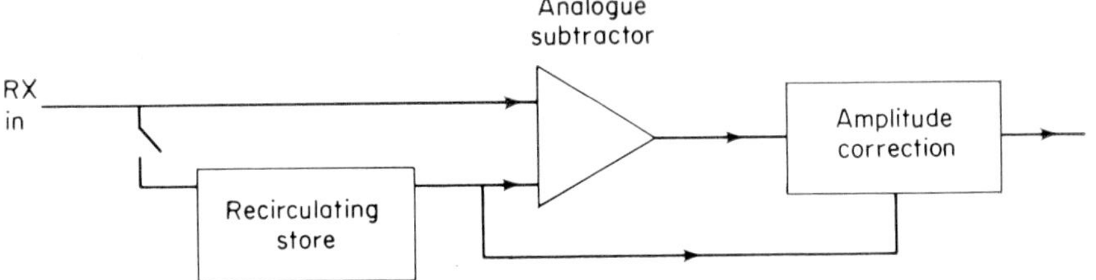

Figure 6.   Signals from large features contain frequency components within the receiver band and must be extracted by means of an analogue subtractor.

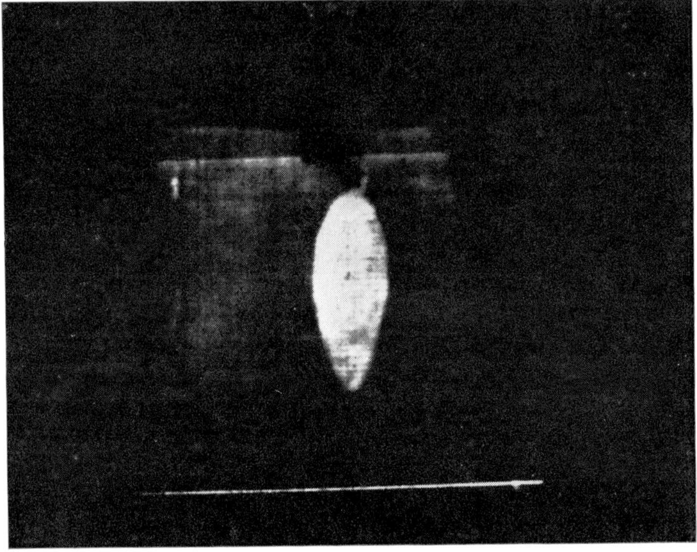

Figure 7.   Reconstruction of a large feature after being extracted by the circuits of figure 6.

3.2. *Large features* (as would typically be of interest to a computerized analyser) have signal frequency components in the same band as the receiver signal and therefore they must be subtracted using a previously stored featureless receiver signal.

Recent advances in fast analogue-to-digital and digital-to-analogue converters and inexpensive digital shift storage allow the receiver signal to be economically stored digitally. In the Ferranti d.c. compensated video receiver, a digital recirculating store continuously supplies the featureless receiver response as a reference. Features are then extracted using on-line analogue subtraction of the featureless receiver signal. Electronic gain compensation is provided by an amplitude modulator controlled by the recirculated featureless receiver signal, the output of the analogue subtractor passing through this modulator (figure 6). The solid nature of the object of figure 5 is then restored (figure 7).

4. Computer application programs

The standard software consists of three parts.

4.1. *The service module* carries out keyboard responses in the control mode, setting up the interface at switch-on and reset, and extracting interface buffer-store data to distribute to other programme modules and to a feature store. This store is continuously loaded for on-line operations but is frozen upon receiving a command from the keyboard or a specific alarm condition. At such a time the feature stored can be viewed either as a teletype-synthesized image or on a v.d.u. with full magnification and image shifting facilities.

4.2. *The feature isolator* takes raw data from the interface and isolates features by associating adjacent zones. Various characteristics of features are accumulated in real time as the activity commences, including perimeter round the feature contour, and feature area. At its lowest level, the feature isolator operates on binary-valued patterns in the zone matrix. For more sophisticated situations it carries out the same operation but on many different contour levels. Feature information for different contours is associated by using positional coordinate information.

The feature isolator therefore carries out a sophisticated real-time process, converting shapes into sets of numbers giving position, size etc. which are available for further computer calculation.

4.3. *The feature filter* takes the data relating to each isolated feature and checks it for acceptance or rejection with templates bearing fault titles related to the manufacturing involved. This program module is generally a special-purpose package to suit the needs of on-line inspection for a particular industry.

Examples of inspection tasks for which this software is suited include:

(*a*) Thickness monitoring of semi-transparent material.
(*b*) Optical-density monitoring of semi-transparent material.
(*c*) The inspection of small components in any orientation passing on a conveyor belt to detect moulding flash, missing pieces etc.
(*d*) The identification of types of fault in sheet finishing industries such as paper, glass and metal.

4.4. *Fault information delay to drive reject marking guns.* This task can be simply accommodated to provide appropriate information packing and down-web delay to suit the application.

4.5. *Optimized cutting schedule.* Software provides drives to on-line cutting tools to maximize the usage of the material output by positioning the required sheet sizes to avoid including faulty areas.

5. Conclusions

The operating system and software so far developed has been built round a particular make of minicomputer. However there are relatively minor differences to be found either between different manufacturers or even with the use of larger and faster computers. Thus the power of the system can be increased readily by the use of more expensive machines. The main direction in which improvement can be gained is in ' traffic density ', i.e. the point at which input must be coarsened to prevent saturation. This is likely to progress favourably anyway because of the inexorable increase in computer speed and reductions in cost of storage.

Laser machines for industry - the breakthrough

J. B. WILLIS

Laser Development, BOC Industrial Power Beams, Daventry NN11 5PJ, England

Abstract. As with any new technology, lasers have taken a considerable time to progress from the euphoria of the initial invention, through the stage where it was shown that lasers could be used in practical applications, to the development of complete laser machines now being used in many different industrial applications.

A brief résumé of the various types of laser that are being used for machining is followed by a description of one company's experience in developing, manufacturing and selling a wide range of industrial laser systems.

1. Industrial lasers

These can be divided into three classes:

Pulsed solid-state lasers of which ruby (chromium-doped aluminium oxide) was the very first example, not only of pulsed, but of any type of laser. Most pulsed systems now depend on the rare earth element neodymium dissolved in either glass, calcium tungstate or yttrium aluminium garnet (YAG). The latter two hosts are crystalline materials that have been found to be particularly suitable for laser work. These lasers can be used in a form which gives short pulses (hundreds of microseconds) for hole drilling or with longer pulses (several milliseconds) for spot welding. They are usually pumped with xenon flash lamps.

Continuous solid-state lasers of which neodymium-doped yttrium aluminium garnet (YAG) is the prime example. This type of laser can produce outputs up to 1 kW, but is usually used in the range up to 50 W. Continuous YAG lasers are pumped by krypton arc discharge lamps and are usually switched at frequencies up to 50 kHz. In this mode they give pulsed outputs with peak powers of several tens of kilowatts. The commonest application is evaporating thin layers or films of various materials.

High power gas lasers which are chiefly represented by CO_2 lasers. The output powers of the commercial forms of these lasers can range from 50 W to 20 kW. The lower powers in this range can be mechanically or electrically pulsed at frequencies up to several kHz. These lasers can be used for cutting, welding and surface hardening a wide range of metals and alloys.

Other types of high power gas lasers are the more powerful of the transverse excited (TEA) pulsed lasers and gas dynamic lasers, which are either chemically, thermally or electrically pumped to give a continuous output. These TEA and gas dynamic lasers are under active development, but are not yet used industrially to any great degree.

2. Laser machines for the wire making industry

Drilling holes in diamonds to make them into wire drawing dies is an expensive, time consuming and laborious process consisting essentially of reciprocating a pointed metal needle in a slurry of diamond powder and lubricant. The process can take up to 24 hours for a single hole so that, to get a reasonable throughput, machines with up to twenty needles or spindles are used which take up space and use large quantities of diamond powder.

Although diamond is transparent, it has been found possible to drill and cut this material with a pulsed ruby laser. Unfortunately, owing to the nature of this laser, the pulses tend to be of a fairly high energy and as often as not caused the stone to shatter. The discovery of neodymium-doped calcium tungstate and YAG gave a laser with a smaller power per pulse, but a frequency of 5 or 10 pulses per second compared to the maximum of 1 per second for ruby. The neodymium laser is used to ' nibble ' the diamond away and when the stone is drilled from both sides gives a bell-shaped profile which is ideal for wire drawing, so reducing the amount of additional polishing necessary.

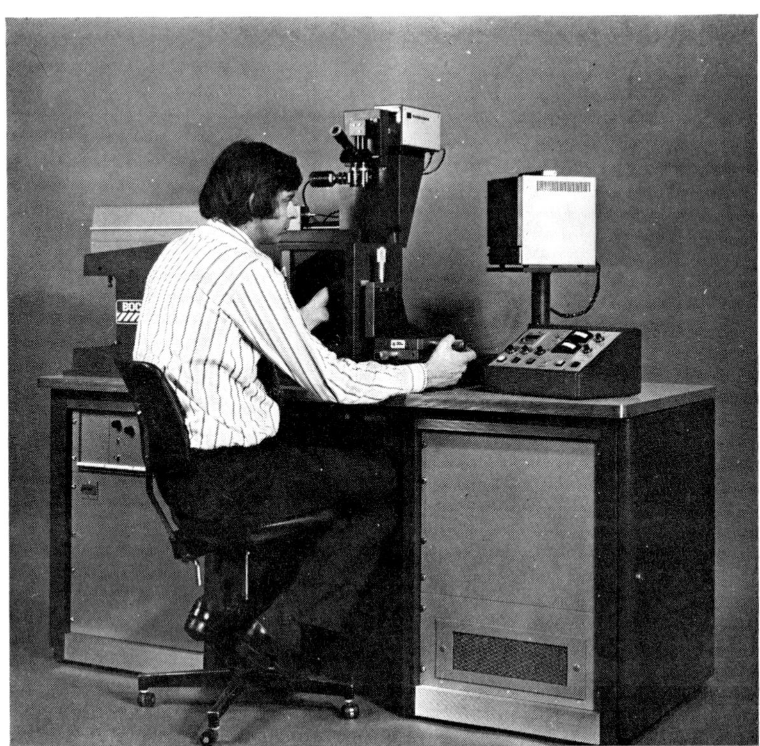

Figure 1.   The BOC diamond-die drilling machine which uses a neodymium/calcium-tungstate laser to drill wire-drawing dies.

The picture in figure 1 shows the BOC diamond-die driller with the laser behind the operator's head. A die is being loaded into the work chamber while the xyz tables are used to position the die at the laser focus. The die is rotated during drilling to make sure the hole is symmetrical and in addition the axis of rotation can be offset from the laser focus so that larger holes can be shaped and trepanned. The whole operation is watched on the CCTV and controlled from the panel on the right which not only enables the operator to set the pulse rate, pulse energy and rotation speed, but also controls the rate at which the pulse energy is increased as the hole deepens.   This ensures that there is minimum damage at the entrance to the hole, but that maximum drilling rate is used in the later stages.

In order to show the cost advantages of the laser system over the traditional method using a twenty-spindle mechanical system, a comparison is made for two sizes of die (Table 1).

Table 1.   Comparison of laser and conventional die drilling

	Laser £	Mechanical £
Basic costs		
Labour plus overheads per hour	2·00	2·00
Spares and servicing per hour	0·35	0·25
0·25 mm diameter dies		
Consumables per hour (Flash tubes, diamond powder, etc.)	0·12	2·09
Die production per hour	12	5
∴ Cost per die	0·20	0·87
Cost saving per die is	0·67	—
Cost saving per machine hour is	8·04	—
1·5 mm diameter dies		
Consumables per hour	0·31	8·62
Die production per hour	4	1·3
Cost per die	0·67	8·36
Cost saving per die	7·69	—
Cost saving per machine hour	30·76	—

If the laser is to give a return of 15 per cent and a depreciation of 20 per cent on its net capital cost of £14 600, then for single-shift working, the cost saving must be at least £2·55 per machine hour.

The economic advantages of laser drilling became abundantly clear in addition to the benefits of cleaner, faster working with more accurate work owing to the extra control given by the CCTV viewing system.

3.   Laser machines for the electronics industry

Very stable and accurate electronic resistors can be made by evaporating nickel-chromium (nichrome) alloy films on to glass or alumina substrates.   These resistors have to be adjusted or trimmed (electrically) by cutting away part of the film while measuring the resistance.   This can be achieved by a spark erosion process, but this tends to give a ragged cut and the electrical discharge which is used seriously interferes with the measurement of resistance.

It was found that these films could be machined with ease using a continuous YAG laser with 2 or 3 W output and mechanically pulsed with a rotating disc. The development of higher power YAG lasers with up to 20 or 30 W outputs and electronic pulsing or Q-switching enabled much thicker films to be cut and thick insulating materials such as alumina ceramics to be scribed by cutting a groove a few thousandths of an inch deep along which the material can be broken.

The discovery of these applications led to the development of special machines which are being widely used in the European electronics industry. A typical machine, shown in figure 2, is the BOC semi-automatic resistor

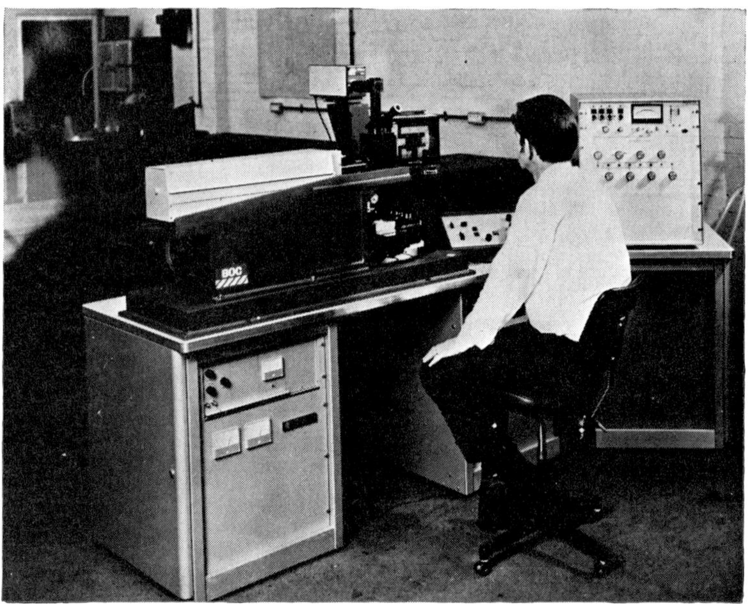

Figure 2.   The BOC semi-automatic resistor trimmer which uses a YAG laser to trim one resistor per substrate.   A typical resistor can be seen on the TV screen and the measuring system is to the right of the operator.

trimmer.   A substrate, usually an alumina plate 0·025 in. thick with a number of resistors and connection strips evaporated or printed on, is loaded into the machine under the cutting lens.   A pair of probes drops across one resistor, connecting it to a measuring system.

When a button is pressed, the laser switches on and the motorized xy tables on which the substrate has been placed moves so that the laser makes a small cut across the resistor.   This cut increases the value of the resistor until it reaches the value set on the measuring system, which then automatically stops the laser and tables.

This system is designed to trim one resistor per substrate at a relatively low capital cost.   The machine can be supplied with a magazine shuttle-feed system to trim up to 5000 resistors per hour at accuracies up to $\pm 0·1$ per cent.

The type of resistor now most often trimmed with this type of equipment is known as thick-film and consists of a mixture of oxides of metals such as ruthenium and bismuth with a glass frit and an organic binder.   This mixture is screen printed on to alumina substrates and fired at a high temperature to form the resistors.   Because of the nature of the process, the resistor is not very accurate and has to be adjusted or trimmed to value.

Thick-film circuits are much cheaper to manufacture than the thin-film circuits described earlier and are of comparable cost to printed-circuit boards with discrete resistors. As a result this type of circuit is being increasingly used in consumer electronics such as colour TV, washing machines and radio, because of the smaller size and greater reliability of these films. The computer industry also uses vast quantities of these circuits.

Because of the high throughput requirements, fully automatic systems have been developed such as the one shown in figure 3, which will trim all the resistors on a substrate. To make contact with all resistors, a multiple probe

Figure 3. The BOC fully automatic resistor trimmer showing the multiprobe assembly under the cutting lens, the computer and control units in the rack, with the teletype to the extreme right of the picture.

assembly is used so that two probes are available for each resistor. The whole system is controlled by a minicomputer and the resistance measurements are made on a programmable unit. When a substrate is loaded, either with a rotating carousel as shown in figure 3, or on a shuttle-feed system, all the probes drop on to the substrate. The resistors are scanned sequentially to check that they are within trimming range and then the resistors are trimmed one at a time as in the semi-automatic system. If any resistor fails to trim to value or is untrimmable then the substrate is automatically rejected and a message is printed on the teletype which is also used to programme and input data to the computer.

The advantages of laser trimming are not entirely economic as all trimming costs are low—for example, trimming thick-film circuits with air-abrasive (a mixture of compressed air and fine alumina powder) costs around £0·002 per trim for 1 per cent accuracy increasing to £0·01 for 0·1 per cent. Laser trimming stays fairly constant at £0·0015 for the semi-automatic system and £0·0008 for the fully automatic: these figures are considered small when compared to the cost of other parts of the thick-film process.

The main advantages lie in the operator acceptance and speed. Laser systems are much preferred to alumina-powder systems which cover everything and everybody with a fine white powder causing contamination and severe wear in all mechanical parts. A single laser system will trim between 4000 and 10 000 resistors per hour, according to type, while an abrasive trimmer will, at best, only cope with a 1000 resistors an hour for 1 per cent accuracy to 100 to 200 an hour when 0·1 per cent is required.

A further resistor trimmer under active development has moving optics, so that the laser beam moves over a stationary substrate, and a digital measuring system to give even higher speeds needed for the very high throughputs necessary for many of today's high volume requirements.

Conventional cylindrical resistors are made by coating thin conducting films on insulated rods. Connections are made by crimping end caps onto the rods and a helical cut is made in the film to increase the value of the resistance by several hundreds or thousands of times. A laser system has been developed, and is now being manufactured, which will adjust these resistors in a fraction of a second compared to many seconds for the high speed graining wheel which is used at present. The laser cuts are also much narrower and more reproducible, which means that much higher magnifications can be achieved to produce high value resistors for the latest very high impedence transistors.

The laser can be fitted to existing spiralling systems or can be supplied as a complete spiralling machine which will produce up to 12 000 resistors per hour.

4. Carbon dioxide lasers in industry

One of the very first industrial applications of CO_2 lasers was to cut dieboards for making cardboard cartons. The original development was carried out in the R and D Laboratories of BOC and subsequently shown in a popular science programme on television. William Thyne Ltd of Edinburgh saw the programme and subsequently co-operated with BOC to develop the Laser Falcon, which is shown in figure 4. The first production system was installed in Thyne's in 1970 and has been in continuous use since then to make, amongst other items, boxes for many brands of Scotch whisky.

An essential part of carton-making is the die used for stamping out the cardboard shapes before they are folded to form the cardboard boxes. The die consists of 0·75 in thick plyboard into which slots are cut to take the cutting blades or ' knives ' which project 0·25 in above the board. The sheets of card are pressed onto the die, resulting in cuts where there are sharp blades and creases where there are blunt blades.

Traditionally cuts are made in a large board by hand using a motorized jig saw or, if very high accuracy is required, the die is made up of a large number of small pieces of the same type of plyboard cut with a rotary saw. These pieces are then locked up with the knives in a forme rather like letterpress printing, to form a multipiece die.

Using a laser, the boards are cut in one piece, as shown in figure 5, in a fraction of the time needed by either of the traditional methods, with a better accuracy than multi-piece dies and without the associated problem of dealing with literally hundreds of pieces of wood.

The Laser Falcon uses a 200 W CO_2 laser with a fixed germanium lens which focuses the laser beam upwards on to the board. The cuts are programmed

Figure 4.   The BOC laser Falcon which uses a CO_2 laser to cut die boards for carton making.   The operator is watching the line follower and below the table is the laser.

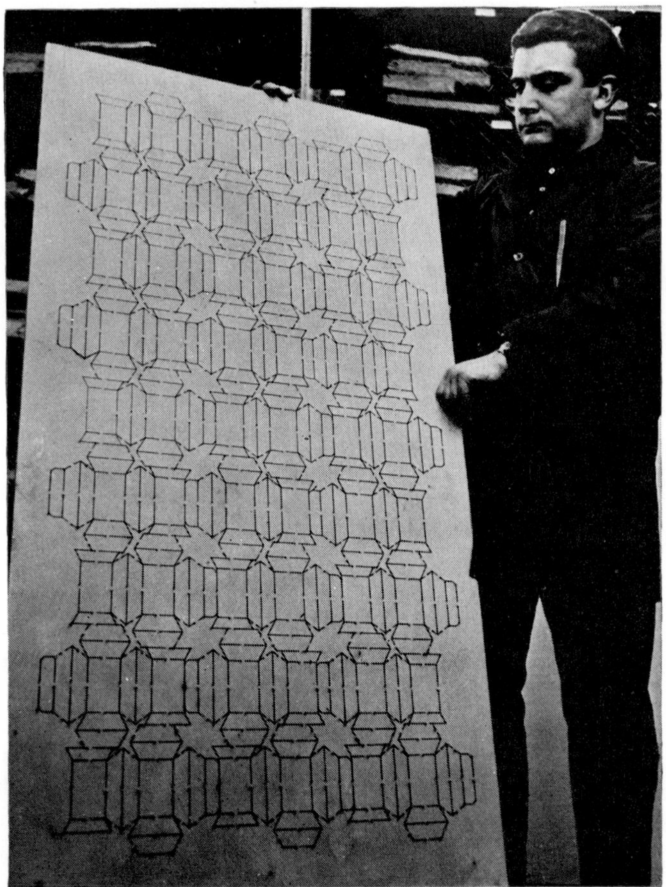

Figure 5.   A single-piece laser-cut carton-die made on a BOC laser Falcon machine.

by moving the board in conjunction with a line following system which tracks over a specially prepared drawing, upon which are superimposed masks with lines of different thickness which tell the laser when to switch on and off and which lines on the drawing to follow.

By using the correct speed, laser power and gas flow (air is normally used as the cutting gas) a uniform parallel-sided cut can be made over the whole board, which can be up to 45 in. by 65 in. in size, while the cut positions are accurate to ± 0.005 in.

The basic machine uses a line-following technique, as this matches with the existing production techniques of many carton manufacturers. Numerical and computer control can just as easily be fitted to the machine, as can higher power CO_2 lasers if higher speeds are necessary.

A study has been made to compare the cost of laser and manual cutting of boards. The results are summarized in Table 2.

Table 2.   Comparison of laser and manual cutting of carton dies

		Laser £	Manual £
Operating costs per hour			
Labour plus overheads		2·00	2·00
Consumable items (gas etc.)		0·23	—
Spares and servicing		0·35	—
	Total	2·58	2·00
Multi-piece dies (average for 19 dies)			
Time in manufacturing and preparation for use		7 hours	36 hours
Total cost per die		18·06	72·00
Cost saving per die		53·94	—
Cost saving per machine hour		7·71	—
Single-piece dies (average for 4 dies)			
Manufacturing time		4·2 hours	39 hours
Total cost per die		10·84	78·00
Cost saving per die		67·16	—
Cost saving per machine hour		15·99	—

For the machine to give a 15 per cent return on capital and allowing for 20 per cent depreciation on the capital cost of £44 000, the total cost saving should be £7·70 per hour. Laser die-cutting machines will pay for themselves on single-shift working, while on the more usual double-shift operation considerable savings are achieved.

At the other end of the scale to the Laser Falcon is the 2 kW CO_2 laser which was developed by the Welding Institute and is now manufactured and marketed by BOC. This machine, which is the most powerful European laser commercially available, has run for many hundreds of hours in the prototype form, the early reliability problems associated with the laser having been overcome. Two of these lasers have been sold to research institutions in the UK and Japan and there are several companies in the UK who are very interested in purchasing complete laser machines for production welding and cutting applications.

There is no doubt that the high power CO_2 laser will become as readily acceptable in industry in applications where its unique properties make it economically and practically viable as has the lower power CO_2 and solid-state lasers.

The 2 kW CO_2 gas laser is capable of cutting up to 12 mm thick stainless steel and deep penetration welding to a depth of 6 mm stainless steel. Typical cutting speed in 2 mm stainless steel is 4 m/min and typical welding speed is $1\frac{1}{2}$ m/min.

High speed cutting has been achieved using gas assistance in a carefully designed nozzle assembly. The gas, usually oxygen, is passed coaxially with the laser beam on to the workpiece. By careful choice of gas flow characteristics, parallel cuts may be achieved. Materials that are subject to fracture or burn damage can be successfully worked with an inert gas jet.

Deep penetration welding can only successfully be achieved above about 1 kW of power. Welds similar to that produced by electron beam can be produced with the 2 kW laser without the necessity of vacuum chambers or the problems of x-ray shielding. Helium is used as the shielding gas for welding.

The main advantages of laser cutting and welding are the small heat input compared to conventional techniques such as plasma; narrow welds and cut widths with small heat affected zones; and repeatability because of the absence of wearing parts such as electrodes or knives.

Applications are being found in the aircraft, automobile, turbine and lamp industries.

5. Safety

Great concern is often expressed about the safety aspects of lasers, especially with the James Bond connotation. There is no doubt that safety is extremely important, but it should be taken in context with other industrial processes. The hazards can be divided into those associated with the electrical supply and those of the laser beam.

High power lasers use high voltages at relatively high currents, for example the 2 kW CO_2 laser uses 20 kV at about 1 A and high energy pulsed lasers use banks of capacitors charged to several kilowatts. All the usual precautions of interlocking doors and panels giving access to high voltage supplies with automatic dumping of capacitors are taken by the laser industry in their equipment to ensure that there is no risk of electrical shock when the equipment is used correctly.

All lasers with powers over 1 mW can cause injury to the eye when looking up the laser beam: neodymium and ruby lasers will damage the retina, while CO_2 lasers will damage the cornea owing to the opacity of the eye at the $10\cdot6\,\mu$m wavelength of this laser. Burn damage to the rest of the body will result from insertion into the beam of a laser with powers of more than a few watts. In all industrial laser systems the beam should be masked so that the operator cannot look along the beam or be able to put any object into its path. Normally all access points have screwed-down covers or are interlocked with fail-safe devices where possible.

It might be worth emphasizing that the radiation from the types of laser discussed here are not ionizing, but are purely light and heat radiations. High intensity light radiation can be emitted during welding and cutting operations with high power pulsed and continuous lasers and the same precautions should be taken as in conventional welding and flame cutting.

6. Economics of industrial laser systems

Once it has been established that a laser can be used to carry out a particular task of cutting, welding, trimming or some other application, one of two requirements must be fulfilled before any attempt is made to develop and manufacture a complete system:

1. Does it enable a process to be carried which is either better than any other method or is not otherwise possible?

2. Does it carry out a process at a lower cost than by any other method taking into account: running costs; labour costs; capital depreciation; production rate; factory space; etc?

For example the diamond die driller is much faster than existing methods and in addition shows cost savings of between £10 and £31 per machine hour, which means that the machine will easily pay for itself in less than a year on even single-shift working. The Laser Falcon shows cost savings of between £7 and £16 per machine hour with a better product and faster production than is possible with hand methods.

The cost savings achieved with the resistor trimmer are not so dramatic, but the rates of production and cleanliness are such that they easily outweigh the existing methods. In addition to greater reliability, high quality small cuts are possible, which give a greater packing density of smaller electronic components.

Reliability is an extremely important factor. Obviously if a machine is twice as efficient as an alternative machine but has a 75% down time, then it is a liability. Laser machines have to be reliable and a lot of effort has to be devoted in the design and development stage to achieve this objective. All the components in the system, laser rods, lamps, mirrors, electrical supplies and so on must be run well within their capability so that the system can run for thousands of hours without more than routine attention such as changing a lamp. Present day systems are designed to be operated and maintained by semi-skilled laboratory staff: it is not necessary to have a qualified scientist with every laser.

7. Conclusions

Lasers are now firmly established in the European electronics, wire-making and carton-forming industries, where it has been demonstrated that well-designed and well-built laser systems have a very important part to play as industrial tools. Reliability and safety are as important as the economic and practical benefits resulting from the use of lasers. In no case have we found a manufacturer who did not feel he had made a great step forward after using laser systems in his plant.

The next stage is the much greater use of high power CO_2 lasers in the engineering industries where we know of many applications where these lasers have already been shown in case studies to have great economic and practical advantages over existing technology. Complete laser systems are being developed in conjunction with several companies in the same way that the diamond die driller, resistor trimmer and Laser Falcon systems were developed with specific companies before being made generally available to industry.

Frequency narrowing and tuning of flashlamp-pumped dye lasers

A. S. NEVILLE

Electro-Photonics Ltd, The Cutts,
Dunmurry, Belfast BT 17 9, Northern Ireland.

Abstract.   The availability of a large number of lasing dyes covering the optical spectrum 330–1200 nm, combined with the frequency tunability of each dye over its broad fluorescence spectrum, makes organic dye lasers the ideal tunable, coherent light source for many applications in various areas of present day research.   Methods of frequency narrowing and tuning of flashlamp-pumped dye lasers are reviewed together with some interesting applications.   In another mode of operation picosecond pulses as short as 2 ps can be reliably and reproducibly obtained from pulsed and c.w. dye lasers.   Direct measurement of both pulse durations and background energy content with high-speed streak cameras have confirmed that pulse durations can be controlled to ± 1 ps and signal-to-background ratios exceed 10^4.   The generation of such pulses from flashlamp-pumped dye lasers is discussed and the present and future applications are reviewed.

1. Introduction

Since 1966 dye lasers have been developed and applied in many laboratories and the scientific literature now contains hundreds of publications dealing with this type of laser.   While, in this paper, it will not be possible to cover all the recent advances and applications, two important properties of dye lasers are discussed and interesting applications in the fields of physics, chemistry and biology reviewed.

Since one of the most useful and attractive properties of dye lasers is the present availability of a large number of lasing dyes covering the optical spectrum from 330 nm to 1200 nm [1], combined with the frequency tunability of each dye over its broad fluorescence spectrum, most emphasis will be placed on frequency control, both narrowing and tuning.   Alternatively the broad output spectra can be mode-locked to produce ultra-short (approximately 2 ps) frequency-tunable high power pulses, with reliability and reproducibility exceeding that obtained with most solid-state and gas lasers.

While previously it was customary to discuss and treat dye lasers according to the method of excitation (laser-pumped, flashlamp-pumped), it would seem more appropriate to consider one method, e.g. flashlamp pumping, and to deal with the basic techniques employed to produce the laser characteristics required for a particular application or experiment.

2. Frequency narrowing and tuning

Laser action in organic dyes results from an inversion of population between a normal Boltzmann distribution of the vibrational population of the first excited singlet state and high energy ($>kT$) vibrational levels of the ground state figure 1 (*a*).   This population inversion occurs after intense irradiation of the dye molecule in its electronic absorption band, for example, by a flashlamp, in a flashlamp-pumped dye laser.   The absorption and fluorescence spectrum of a typical large organic molecule is shown in figure 1 (*b*).   The fluorescence band can be seen to stretch over several tens of nanometers.   This broad fluorescence stands in contrast to the narrow, sharp emission of most other lasers and leads to the tunability of the laser emission that makes organic dye lasers so important.

A schematic of a flashlamp-pumped dye laser is shown in figure 2.   The optical resonator is formed by two mirrors (a, b).   Energy is delivered to the

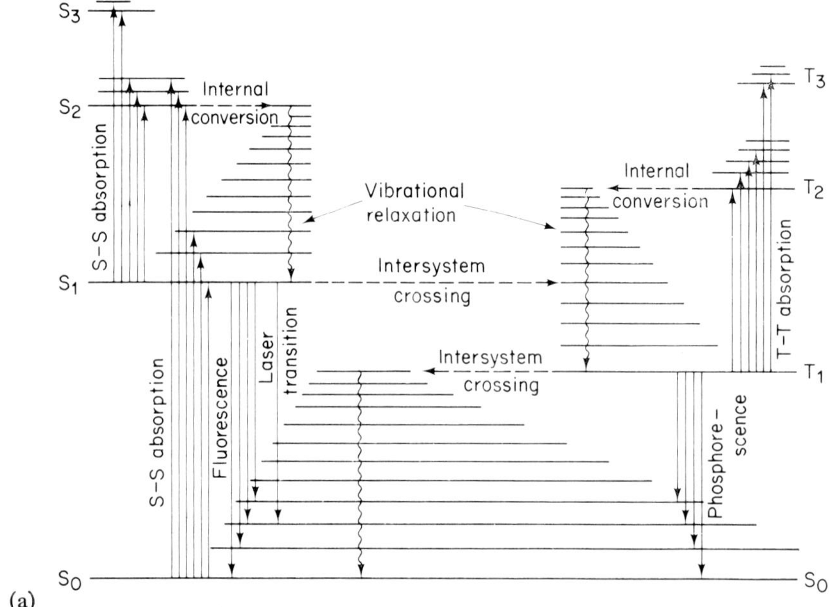

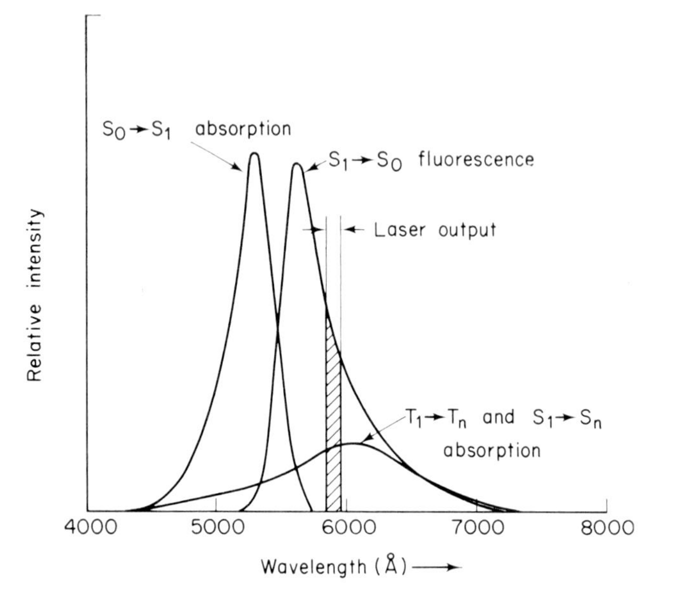

Figure 1.   (*a*) Electronic levels of dye molecules.   (*b*) Absorption and fluorescence spectra of Rhodamine 6G.

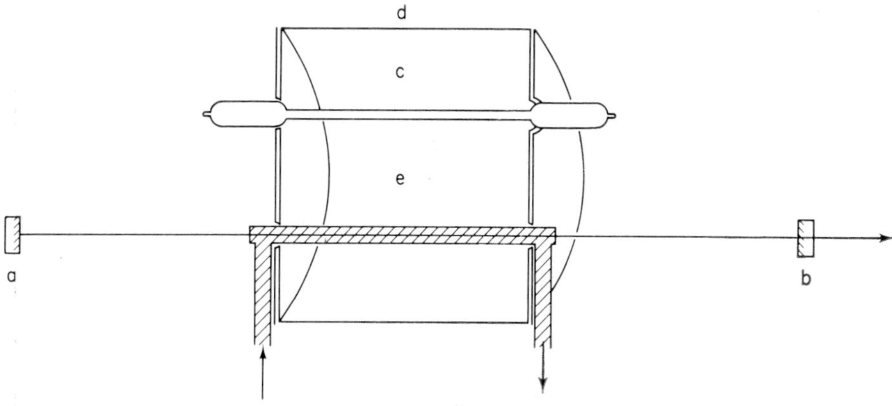

a. Dielectric coated max. reflector
b. Dielectric coated output mirror
c. Xenon flashlamp
d. Elliptical cavity
e. Cell containing flowing dye

Figure 2.   Optical resonator of flashlamp-pumped dye laser.

dye from the flashlamp (c): efficient collection of this energy is aided by the elliptical reflector (d) that surrounds the flashlamp and the dye cell.   The dye is circulated through the optical cell (e) to minimize thermal distortions of the optical media and the effect of photochemical decomposition of the dye.

The system shown would emit laser radiation in a continuum approximately 6–10 nm wide.   Frequency narrowing and tuning are achieved by employing dispersive elements inside the laser cavity, e.g. diffraction grating, prism, Fabry-Perot interferometer or birefringent filter.   It is our experience that when diffraction gratings are employed for tuning high-energy flashlamp-pumped dye lasers, the thermal refraction effects in the dyes lead to the appearance of satellite lines in the output spectra.   Depending on the output energies, anomalous spectral structure can also appear with multiple-prism tuning systems.   However Schäfer and Muller [2] have reported a six-prism tunable-ring laser producing a clean spectral line of 50 pm bandwidth.   For frequency narrowing and tuning high energy dye lasers (output energy approximately 0·5 J) we employ optically contacted thermally compensated Fabry-Perot interferometers, first developed for this application by Bradley and co-workers [3].   The Fabry-Perot etalon has an angular dispersion greatly superior to that of a diffraction grating and frequency selection is less sensitive to increased beam divergence arising from thermal gradients in a flashlamp-pumped dye laser.   Inserting a single narrow gap interferometer (gaps 5–6 μm) into the optical resonator we obtain output energies of 0·5 J, bandwidth 0·3 nm. Further spectral narrowing can be achieved by the addition of a second wider-gap Fabry-Perot etalon into the optical resonator: bandwidths with two etalons in cascade are 0·01 nm.   Figure 3 shows a high-energy dye laser with double Fabry-Perot interferometer tuning.

Figure 4 shows the tuned spectra of a Rhodamine 6G dye laser tuned with a single narrow-gap Fabry-Perot interferometer rotated through a total angle of 30°, corresponding to nearly two free spectral ranges.   (The top two spectra

A. S. Neville

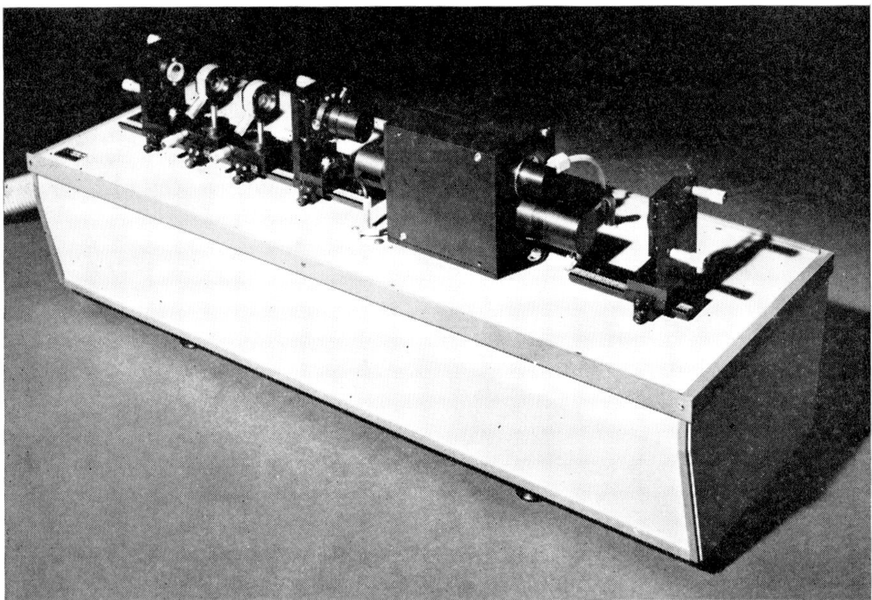

Figure 3. High energy flashlamp-pumped dye laser with double Fabry-Perot interferometer tuning.

are a mercury lamp and the untuned laser respectively. Untuned laser bandwidth is 6·5 nm at 595 nm. Tuned bandwidth is 0·3 nm). Figure 5 shows a typical energy-versus-wavelength tuning curve for a high-energy flashlamp-pumped dye laser: the tuning range covers the visible spectrum over 435–700 nm. Figure 6 shows the spectral output of this laser with Rhodamine 6G tuned by two interferometers in cascade. The laser line, here shown to be narrowed to the resolution limit of the spectrograph, is 0·01 nm. The narrowed line is shown to be tuned through 1·8 nm the f.s.r. of the second etalon. The energy in this narrow laser line is approximately 250 mJ corresponding to a pulse of peak power 200 kW. For applications requiring extreme spectral purity of laser output, three etalons can be used in cascade. Gale [4] achieved single transverse and longitudinal mode operation with a flashlamp-pumped dye laser using three Fabry-Perot etalons of plate separations $10 \,\mu\text{m}$, $500 \,\mu\text{m}$ and 5 mm,: a spectral width of 4 MHz was achieved. Single-transverse, single-longitudinal mode operation of a c.w. dye laser has also been achieved by Hercher and Pike [5], with a bandwidth of 35 MHz, while Hansch [6] has succeeded in narrowing to 300 MHz a Rhodamine dye laser, repetitively pumped by a nitrogen laser, by employing the combination of a beam expanding telescope, tilted etalon and Littrow mounted diffraction grating as cavity elements.

As mentioned previously multi-prism tuning systems provide convenient tuning elements when output energies are low. We have successfully incorporated a three-Brewster-angled-prism tuning element in a low energy (≈ 10 mJ) high repetition rate (15 Hz) flashlamp-pumped dye laser. The optical layout of the resonator is shown in figure 7: frequency tuning is done by utilizing the dispersing power of the prisms and tilting the 100% mirror. With this arrangement, linewidths of 0.012-0.02 nm are achieved over the wavelength range 440 - 700 nm.

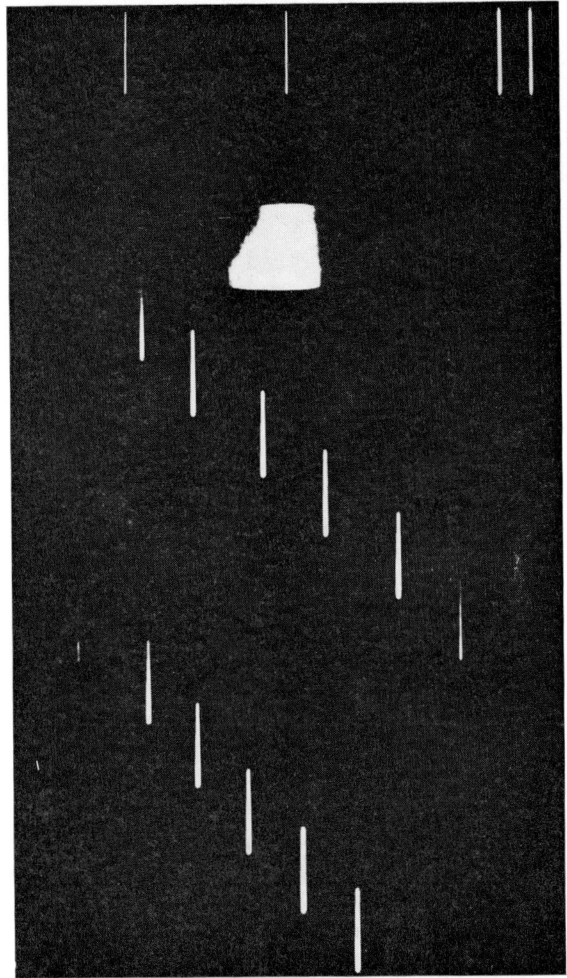

Figure 4. Successive exposures of spectra (1 m Ebert spectrograph) of Rhodamine 6G dye laser tuned with single narrow-gap Fabry-Perot interferometer. By rotation of the interferometer through approximately 30°, the free spectral range is scanned through twice.

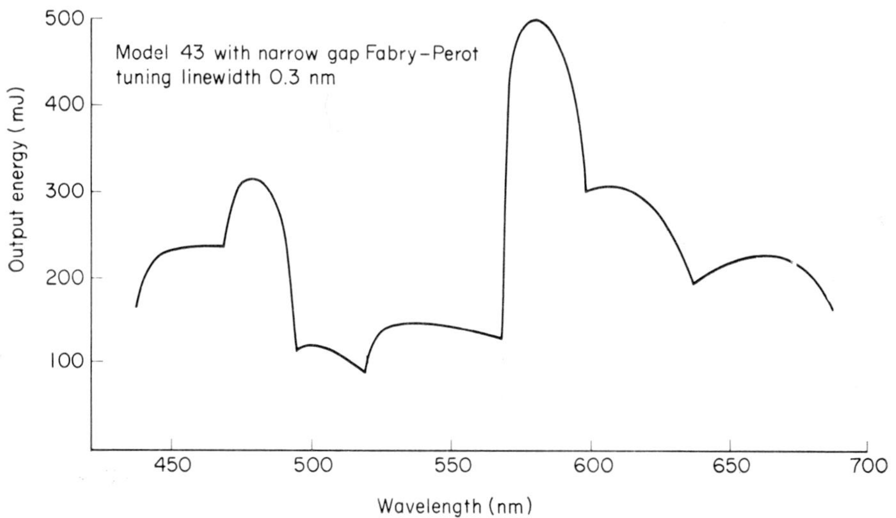

Figure 5. Energy-versus-wavelength tuning curve of high-energy dye laser system.

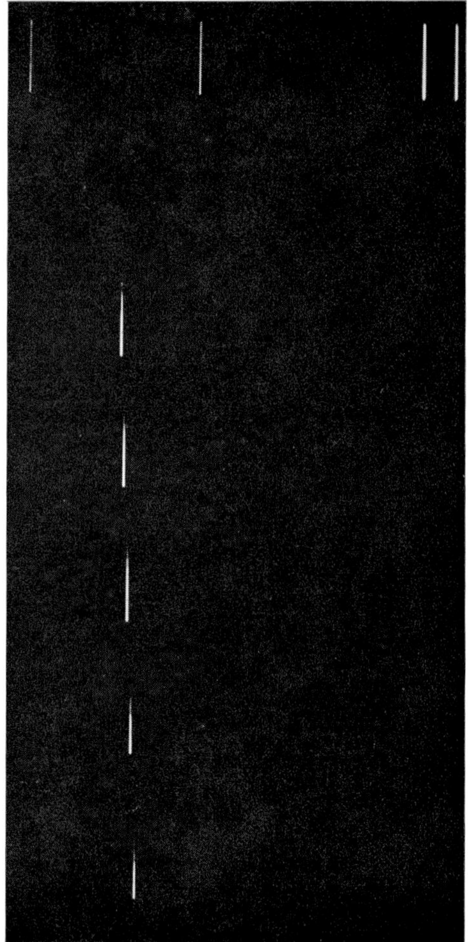

Figure 6. Spectral output of a Rhodamine 6G dye laser tuned by two interferometers in cascade. The laser line, here shown to be narrowed to the resolution limit of the spectrograph, was measured to be 0·01 nm, using a 5 mm gap interferometer. The narrowed line is shown tuned through 1·8 nm.

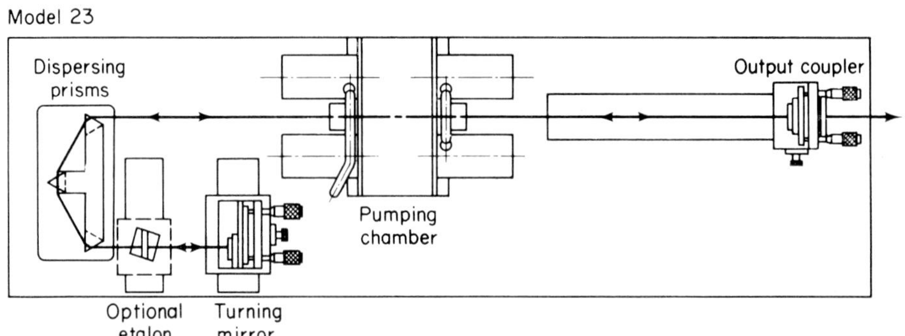

Figure 7. Optical resonator of a high-repetition-rate dye laser system with prism tuning.

The energy-versus-wavelength tuning range for this laser is shown in figure 8: this is achieved with a range of seven laser dyes. Additional frequency narrowing

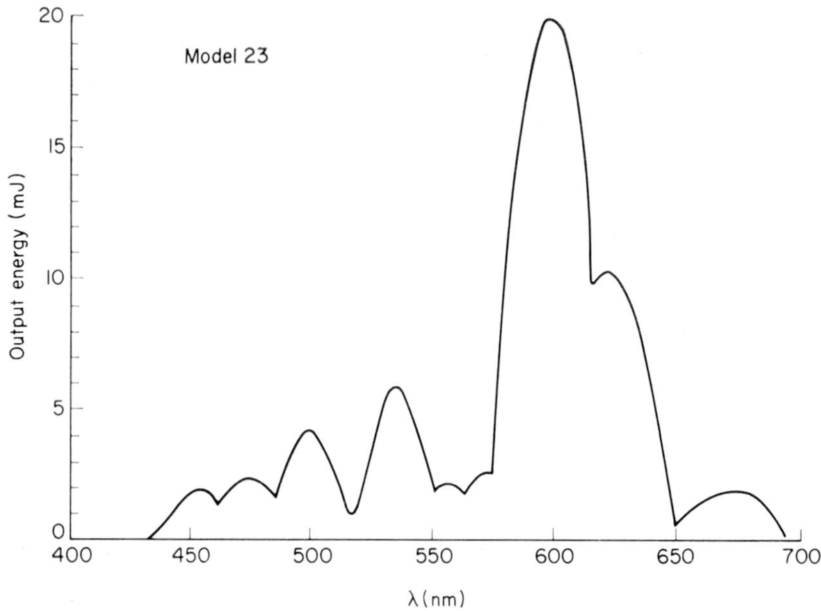

Figure 8.   Energy-versus-wavelength tuning curve of high-repetition rate laser.

can be obtained by inserting a suitable Fabry-Perot etalon into the resonator and linewidths of <0.004 nm can be achieved.   By frequency doubling, with a range of non-linear crystals, tunable radiation in the wavelength range 265–350 nm can also be obtained with this laser system.   The diffraction limited, highly polarized fundamental laser output lends itself to very efficient second harmonic generation and peak pulse powers of up to 1 kW are attainable at 300 nm, a wavelength region of particular interest, since it coincides with the absorption spectra of the oxides of common pollutants such as sulphur and nitrogen. Present development on frequency tuning and narrowing includes the use of birefringent filters for flashlamp-pumped dye lasers.   A three-element birefringent filter provides narrowing to 0.1–0.2 nm and a five element filter to 0.015 nm.   Tuning can be over the entire visible and is accomplished by manual rotation of the tuning element or automatically when linked to a stepping motor. This will be extremely useful in applications which, for example, require the laser line to be scanned across an absorption line in small increments of <1 nm.

Early applications for frequency-tunable dye lasers included, selective excitation of atomic levels [7], self-induced transparency [8], resonance scattering from atmospheric sodium [9], and contour holography [10].   Recent interesting applications include the saturation spectroscopy technique, which may be used to study molecular transitions with a resolution which is within the linewidth of the laser.   This involves saturating the species with the laser beam and then observing the dip (lessening) in absorption as the linewidth of the beam is scanned, at the actual species absorption line.   Resolution within the Doppler width may be achieved.   In an extension of this technique the Doppler broadening can be removed and true line shapes be measured.   The experimental set up

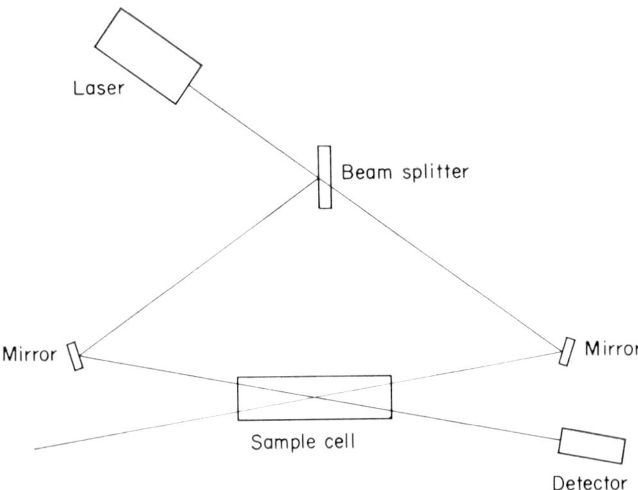

Figure 9.   Optical layout for saturation spectroscopy for true line shape measurements,
i.e. removal of Doppler broadening factors.

is shown in figure 9: at the crossover point only molecules with zero velocity are being sampled by both the pump beam and probe beam.   Modulation of the pump beam leads to a modulation of molecular population and results in modulation of the probe beam only when zero-velocity components are being sampled. Hansch and Schawlow [11] have used this technique to measure atomic sodium spectra when Doppler broadening is removed.   Strictly speaking, saturation spectroscopy is not a new technique, but in the past it has been restricted to the study of molecular and atomic transitions that were accidently coincident with laser emissions.   The use of tunable dye lasers removes the restrictions of accidental coincidence and permits the study of any transition from 330 nm to 1200 nm.   The tunability of dye lasers is of course directly applicable to fluorescence studies.   One interesting application has been the development of an airborne laser fluorosensor for the detection of algae in the sea [12].   The system transmits dye laser radiation at 590 nm and monitors fluorescent emission at 685 nm from chlorophyll-bearing microorganisms.   Applications in the field of biology and medicine include holographic microscopy [13].   In biological experiments a specimen can be fixed in time and its 3-dimensional structure can be recorded on a holographic film and reconstructed at will.   The advantage of using a tunable dye laser for the method is that double-exposure holograms can be made at two different wavelengths, and also unwanted interference and speckle patterns, due to the numerous optical surface variations found in a typical microscope, are eliminated.   Using a tunable dye laser it is possible to destroy harmful bacteria in a given medium without affecting other useful bacteria [13]. Each bacteria class has a unique absorption band and therefore selective destruction is possible.   Applications include control of bacteria in water, milk and dairy goods.   The process is made more powerful by using frequency doubling of the visible dye-laser radiation, since nucleic acids have a strong absorption band around 265 nm: this leads to a 3×10^4 advantage over a visible output at 600 nm.

3. Mode-locking and the generation of frequency tunable picosecond pulses

The broad output spectra of dye lasers can be mode-locked to produce ultra-short frequency-tunable high-power pulses with many applications in the research of transient optical effects, fast reactions, etc. Flashlamp-pumped dye lasers were first mode-locked in 1969 [14] by employing a saturable absorber dye solution in the laser resonator. Figure 10 shows a flashlamp-pumped mode-locked dye-laser resonator. The saturable absorber is contained in a thin dye

Model 33

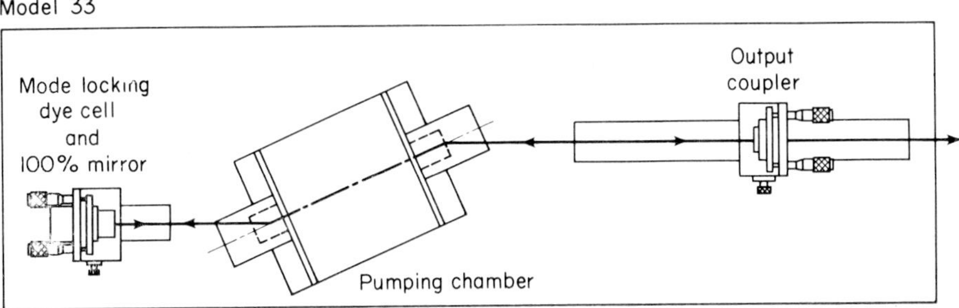

Figure 10. Optical resonator of mode-locked dye laser system.

cell and is in contact with the 100% laser mirror. This position was found to be the optimum for good mode-locking by Bradley and co-workers [14]. By employing different polymethine dyes as saturable absorbers and a narrow gap Fabry-Perot interferometer, picosecond pulses, frequency tunable from 585 nm to 700 nm can be generated with the three lasing dyes: Rhodamine 6G, Rhodamine B and Cresyl-Violet [15]. The trains of mode-locked picosecond pulses can be generated with great reliability and reproducibility. Direct measurements [16–18] of the durations of individual pulses by an electron-optical streak camera, with time resolution <5 ps, have shown these pulses to be reproducible to ±1 ps. With this direct-measurement technique, it has also been possible to show [19] that there is less than 3 per cent of the total energy outside the picosecond pulses, and that the peak power exceeds 20 MW with a signal-to-background ratio of better that 10^4. When operated untuned or tuned, but well above threshold, pulse widths are of duration 5 ps with laser spectral bandwidths of 3–10 nm: these bandwidths are much greater than the Fourier transform limit set by the pulse durations. This frequency broadening of the picosecond pulses is caused by self-phase modulation effects in the dye medium. However, operating near threshold and tuning with a narrow gap Fabry-Perot interferometer, the pulse-bandwidth/duration discrepancy can be removed and picosecond pulses of transform-limited durations ($\Delta t \approx 3$ ps, $\Delta t \, \Delta \nu \approx 0.5$) can be obtained [20]. The tuned output spectra obtained from a Rhodamine 6G mode-locked dye laser delivering picosecond pulses of transform-limited durations is shown in figure 11.

Continuous-wave dye lasers [21] have been mode-locked by employing intracavity modulators [22], to produce pulses [23] of durations ≈ 55 ps. Passive mode-locking of Rhodamine 6G c.w. lasers has also been obtained with the same mode-locking dye (DODCI) employed with pulsed lasers [24]. The availability of mode-locked dye lasers has permitted advances in time resolution of streak cameras. Bradley and co-workers [16–18], pioneers of the Photochron I streak

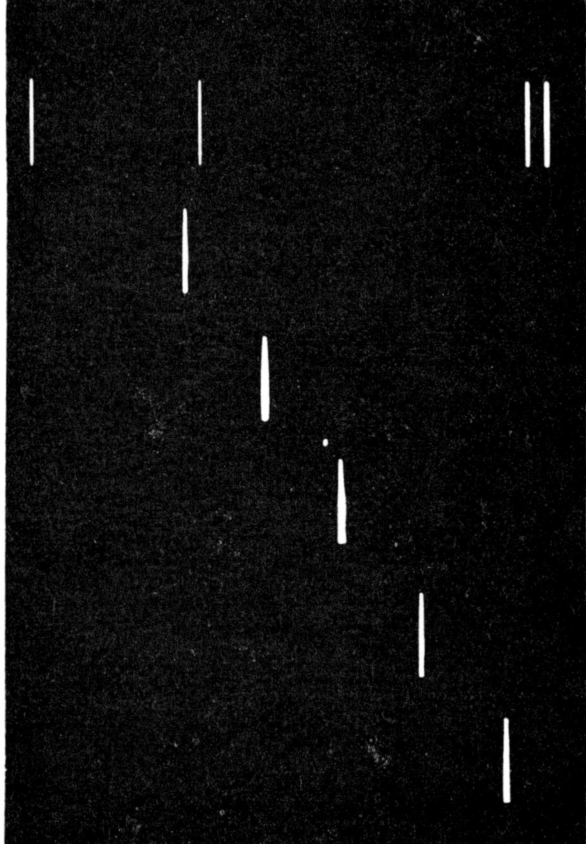

Figure 11. Spectra showing tuning of transform-limited picosecond pulses. Mercury calibration at top. Plate of 1 m spectrograph was moved vertically as laser was tuned from 603 nm to 625 nm.

camera with time resolution less than 3 ps, have recently developed a new streak camera, Photochron II, capable of subpicosecond resolution. They used a mode-locked dye laser pulse to generate a subpicosecond Raman laser pulse in ethanol, which in turn was used to test the subpicosecond resolution capabilities of the new camera system. The mode-locked dye laser and the ultra-fast image-converter streak camera have been used in tandem to form a diagnostic tool in many areas of research. For example Sigel *et al.* [25] has used one of our mode-locked dye lasers together with our ICC 512 image-converter streak camera for an experimental investigation of shock waves in transparent solids driven by laser ablation. Shock waves in plane solid hydrogen and lucite targets were investigated. They were induced by focusing a 20 J/ns Nd-glass laser pulse onto the surface of the target. The time evolution of the shock wave was measured by means of the streak camera and a dye laser as a light source during and after the laser pulse using shadowgraph and Schlieren methods. By mode-locking the dye laser, a sequence of framing pictures were also obtained, which showed the spatial growth of the shock wave with time. The experimental arrangement is shown in figure 12. Using this set-up, important data on the pressure behind a laser-driven shock wave was obtained. These results provided further information for the large world-wide research effort towards laser fusion. In another novel application Coutts *et al.* [26] constructed a high-speed radiation camera incorporating a mode-locked dye laser. High-speed radiation cameras typically employ fast scintillators and framing cameras constructed of electronically shuttered image-intensifier arrays.

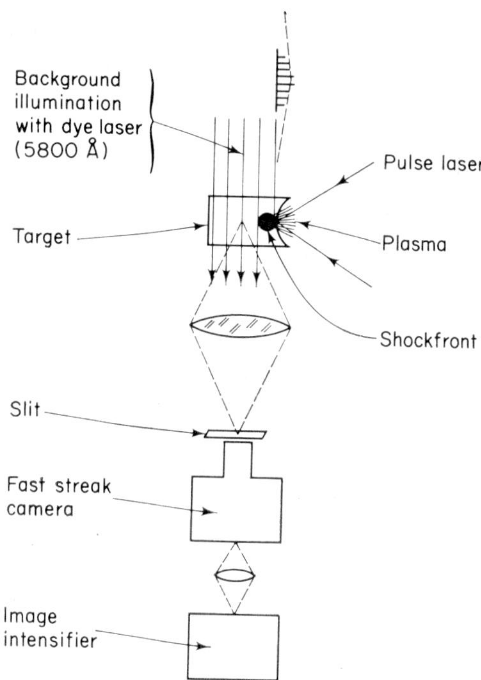

Figure 12. Experimental arrangement for observation of the laser-induced compression wave in a solid target. (Courtesy of Max Planck Institute Garching, Munich).

These systems can take high-speed photographs of pulsed radiation sources, but their meaningful time resolution is limited to 1·7 ns by fluorescence lifetime of scintillator material and by the gating of commercial image intensifiers. To avoid the limitations, the authors developed an ultra-high-speed camera which uses the photoabsorption phenomena associated with hydrated electrons and the picosecond pulses obtained from our mode-locked dye laser. Acidic aqueous solutions function as absorption mechanisms that convert rapidly changing radiation patterns into corresponding optical absorption patterns. Separate, time-resolved, high resolution images of these transient absorption patterns can be transferred to photographic film with short laser pulses. The effective exposure time for the transferred image is limited by the absorption-centre lifetime and the width of the sampling laser pulse. Since the mode-locked laser pulses are approximately 5 ps and the absorption centre lifetime is less than 20 ps, the potential time resolution for this system is approximately 20 ps. This is far superior to existing scintillator/framing-camera systems.

The applications of mode-locked dye lasers cited above are to a certain extent novel applications. The availability of frequency-tunable picosecond pulses is most important in the study of rapid phenomena in materials. Picosecond pulses provide the key to the understanding of the most fundamental processes in materials physics and chemistry. Since particle-collision rates within materials are rapid, the picosecond timescale is the appropriate gauge for a large number of energy-transfer processes in substances. In the techniques used to study the time development of rapid phenomena, the material is first disrupted from equilibrium by optical excitation with short intense picosecond light pulses, and then the rate of return is studied by one of a few basic techniques, the optical Kerr gate, a probe-beam technique and the streak camera.

Using picosecond pulses and these techniques, studies of rapid phenomena are being undertaken in the fields of biophysics, plasma physics, solid- and liquid-state physics, chemistry and photochemistry. Research areas include the study of photosynthetic photovisual processes [27], phonon exciton and polarization lifetimes [28], measurements of ' dephasing time ', which is the duration for molecules that are vibrating together coherently to go out of phase relative to one another [29], rate of formation of a charge-transfer complex in liquids [30]. It is obvious that in the future, mode-locked picosecond pulses will play an increasing role in the study of the fundamental processes of nature.

4. References

[1] BASS, M., DEUTSCH, T. F., and WEBER, M. J., 1971, in *Lasers* **3**.
[2] SCHAFER, F. P., and MULLER, H., 1971, *Opt. Commun.*, **2**, 407.
[3] BRADLEY, D. J., CAUGHEY, W. G. I., VUKUSIC, J. I., 1971, *Opt. Commun.*, **4**, 150.
[4] GALE, G. M., 1973, *Opt. Commun.*, **7**, 86.
[5] HERCHER, M., PIKE, H. A., 1971, *Opt. Commun.*, **3**, 65.
[6] HANSCH, T. W., 1972, *Appl. Optics.*, **4**, 895.
[7] BRADLEY, D. J., GALE, G. M., SMITH, P. D., 1970, *Journal of Physics B. Atomic & Molecular Physics*, **3**, 11.
[8] BRADLEY, D. J., GALE, G. M., SMITH, P. D., 1970, *Nature* No. 5234, 719.
[9] SANFORD, M. C. W., GIBSON, A. J., 1970, *J. atmosp. terr. Physics*, **32**, 1423.
[10] SCHMIDT, W., VOGEL, A., PREUSSLER, D., 1973, *Appl. Phys.*, **1**, 103.
[11] HANSCH, T. W., SHAKIN, I. S., SCHAWLOW, A. L., 1971, *Phys. Rev. Lett.*, **27**, 407.
[12] KIM H. H., July 1973, *Appl. Optics*, **12**, 1454.
[13] HARRIS, 1973, *American Laboratory*, 93.
[14] BRADLEY, D. J., O'NEILL, F., 1969, *Journal of Opto-Electronics*, **1**, 69.
[15] ARTHURS, E. G., BRADLEY, D. J., RODDIE, A. G., 1972, *Appl. Phys. Lett.*, **20**, 125.
[16] BRADLEY, D. J., 1973, U.K. Patent 1329977 (1973), US Patent 3761614 (1973).
[17] BRADLEY, D. J., LIDDY, B., SLEAT, W. E., 1971, *Opt. Commun.*, **2**, 391.
[18] BRADLEY, D. J., LIDDY, B., SIBBET, W., SLEAT, W. E., 1972, *Appl. Phys. Lett.*, **20**, 219.
[19] BRADLEY, D. J., LIDDY, B., RODDIE, A. G., SIBBET, W., SLEAT, W. E., 1971, *Opt. Commun.*, **3**, 426.
[20] ARTHURS, E. G., BRADLEY, D. J., RODDIE, A. G., 1971, *Appl. Phys. Lett.*, **19**, 480.
[21] PETERSON, O. G., TUCCIO, S. A., SNAVELY, B. B., 1970, *Appl. Phys. Lett.*, **17**, 245.
[22] KUIZENGA, D. J., 1971, *Appl. Phys. Lett.*, **19**, 260.
[23] DIENES, A., IPPEN, E. P., SHANK, C. V., 1971, *Appl. Phys. Lett.*, **19**, 258.
[24] SHANK, C. V., IPPEN, E. P., May 1972, VII International Quantum Electronics Conference, Digest of Technical Papers, p. 7.
[25] VAN KESSEL, C. G. M., SIGEL, R., 1973, *Applied Physics Society Bull. II*, **18**, 1316.
[26] COULTS, G. W., *et al.* May 1975, I.E.E.E./O.S.A. Conference on Laser Engineering and applications. Digest of Technical Papers.
[27] RENTZEPIS, P. M., 1970, *Science*, **169**, 17.
[28] ALANO, R. D., SHAPIRO, S. L., June 1973, *Scientific American*, 42.
[29] LAUBEREAU, A., KAISER, W., 1974, *Opto-Electronics*, **6**, 1.
[30] EISENTHAL, K. B., 1975, *Acc. Chemistry Research*, **8**, 118.

Special purpose electro-optical equipment

L. R. BAKER

Sira Institute Ltd, South Hill, Chislehurst, Kent, England.

Abstract. Manufacturers and users of electro-optical systems occasionally require special purpose equipment which can often be supplied most efficiently by instrument contract research and development organizations. Typical examples of instruments supplied to customers under contract for the automatic inspection of strip products, such as steel, and plastics, components such as lenses, and blades used in cutting and wiping are described.

1. Introduction

In a new and rapidly expanding technology, such as electro-optics, it would be unwise for a small firm to risk the establishment of a special research and development facility to develop a new product when the market need for that product may have changed in the time taken to establish the facility. In this situation it is preferable to hire the research and development facilities provided by other well-established organizations specializing in particular fields of technology and a number of these exist in Britain.

The service may be required by either a manufacturer or a user of electro-optical systems. The manufacturer may need to improve the performance of his current range of instruments or to expand his present range by adding new instruments. The user on the other hand may need a new instrument not available commercially to help solve a difficult problem or as an aid to exploratory development.

The contract research and development organization in each case functions as an extension of the research and development facilities of the company seeking the service. The work normally proceeds through a number of stages of varying duration depending on the complexity of the problem. The first stage is to prepare a technical brief and performance specification. This stage requires close liaison with the customer to determine exact needs and to select the preferred solution from a number of alternatives.

At the end of this stage sufficient information has been generated and reported back to the customer to enable a price to be quoted for the design of the prototype instrument. On completion of design, a price can be quoted for the manufacture of the prototype. The prototype is usually delivered with engineering drawings and detailed operating manual to enable the customer to gain industrial operational experience and to decide the most appropriate means for subsequent manufacture.

The design, development and manufacture of an electro-optical instrument requires a multi-disciplinary approach. Particular care has to be taken in choosing optical and mechanical designs which can be economically manufactured and arrangements which are ergonomically adequate as well as pleasing in appearance. There is always the need moreover to resist depending on

components which may not continue in production or whose performance may
be unreliable.

By way of example this paper describes a selection of special purpose electro-
optical equipment which has been developed on behalf of particular customers,
for the inspection of strip products, components and blade edges.

2. Inspection of strip products

Many strip products manufactured from such materials as steel, aluminium,
plastics and paper are currently inspected visually during manufacture. The
defects observed may merely detract from the appearance of the product, as
would scratches or rolled-in dirt on stainless steel, or they may be functionally
undesirable, such as faults in protective tinplate. At throughput speeds of
perhaps 1000 m/min, visual inspection is quite impossible. Several optical
scanning systems are now in use on strip material.

A new, fast, laser scanning system for inspection of plain strip material is
currently under development. The optical layout, which has been designed to
allow material as wide as 1·5 m to be inspected with a single optical head, is
shown in figure 1. The beam from the laser is converted from a circular to a
rectangular cross-section and is then deflected by a multi-faceted mirror drum
which rotates at high speed. The beam is then incident on a long plane mirror
which deflects on to the surface under inspection via a long, curved mirror a
little longer than the width of the material to be examined. A photomultiplier is

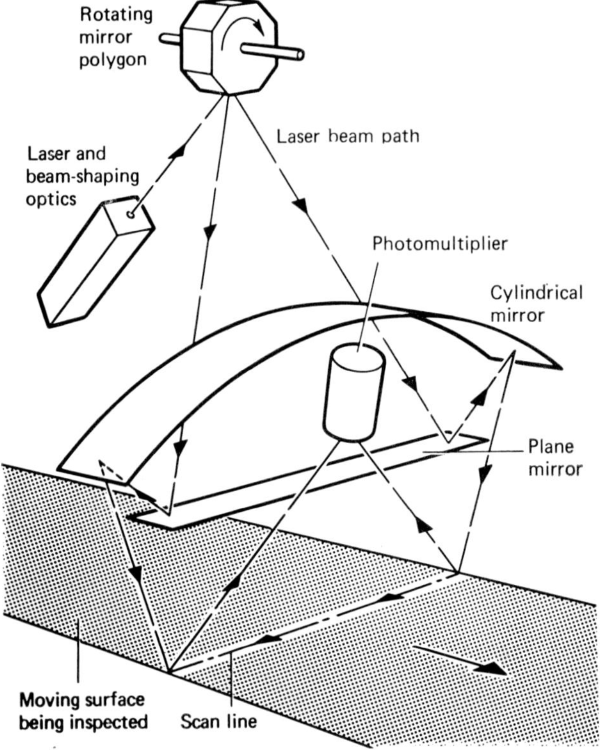

Figure 1.   Flying spot laser scanner for metal surface inspection.

positioned at the centre of curvature of the curved mirror and picks up the light reflected specularly from the inspected material. The laser beam is scanned across the strip material by the rotation of the mirror drum. To give 100% coverage of the material the scan in the orthogonal direction is provided by the continuous movement of the material itself.

The electronic signal generated by the photomultiplier is in the form of a square-wave pedestal whose leading and trailing edges correspond to the strip edges. Surface defects cause localized changes in the level of this signal and by suitable signal processing these impulses can be separated from the much slower changes in reflection arising for example from the variations in angle of incidence of the beam. By the use of a comparator, circuit pulses known to arise from defects of interest can be displayed or counted. Figure 2 shows a c.r.t. display of a defective area of tinplate as seen by the instrument, together with a photograph of the material itself.

Tests with a version of the equipment covering 550 mm wide material have shown that it is possible to detect defects equivalent to a 0·1 mm black spot on good quality stainless steel. With reduced sensitivity, strip moving as fast as 4000 m/min can be 100% inspected.

Another labour-intensive area where inspection of strip products is required is in the printing industry. Although automatic register monitoring devices and stroboscopic viewers are commonly used as operator aids on the printing press there is at present no means of monitoring overall print quality for such defects as lateral misregister of colours, ink spots, missing print and half-tone colour variations. An instrument is currently being developed which aims to carry out 100% inspection of printed material for the common printing faults. The system operates by forming the Fourier transform of the distribution of intensity at one spatial frequency along a strip of the pattern lying at right angles

Figure 2(*a*).   Severe lamination defect on sheet steel.

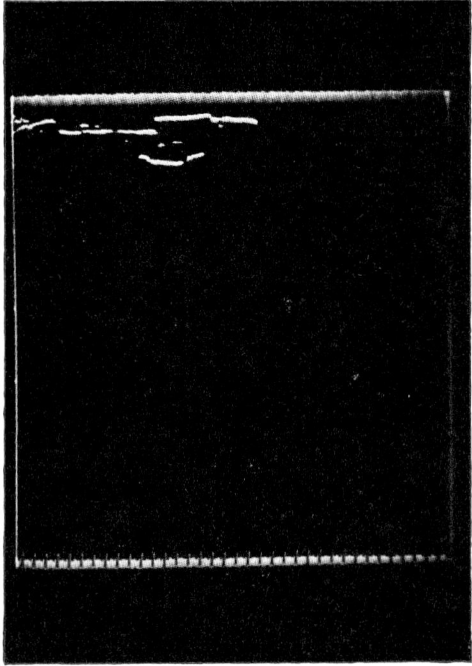

Figure 2(*b*).   Defect in 2(*a*) as instrument c.r.t. display.

to the direction of movement of the printed material, which may for example be a reel of labels or individual sheets of cartons.

A narrow strip across the image of the illuminated printed pattern is selected and focused onto a multiple-aperture grid, behind which is placed a condenser and photodetector (figure 3).   The grid is in the form of a radial grating which rotates at a constant speed and so a constant frequency signal is given by the

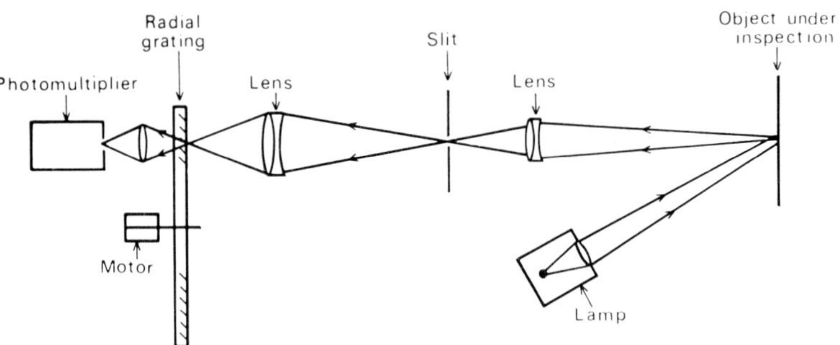

Figure 3.   Optical system for inspection of printed patterns.

detector.   The amplitude of modulation of this signal after filtering is the Fourier component present at the spatial frequency defined by the reciprocal of the line spacing of the grating.   The mean reflectivity is recorded as the zero order or unmodulated component of the output signal.   After analogue-to-digital conversion, the output data are processed by a small, fast digital data processor which compares by cross-correlation stored data derived from prior inspection of a reference sample with data produced by each test sample in turn. Additional units required are a clock unit to provide a continuous train of pulses linked to product speed, which operates the analogue-to-digital converter, and a marker unit to trigger the digital processor when comparison is to be made. The tolerance level of correlation is set to a level acceptable to the customer.

Applications of the technique under investigation so far include high-speed inspection of cartons and labels on-press and off-press at the sorting stage. Figure 4 shows the prototype inspection head in use on a label-reeling machine.

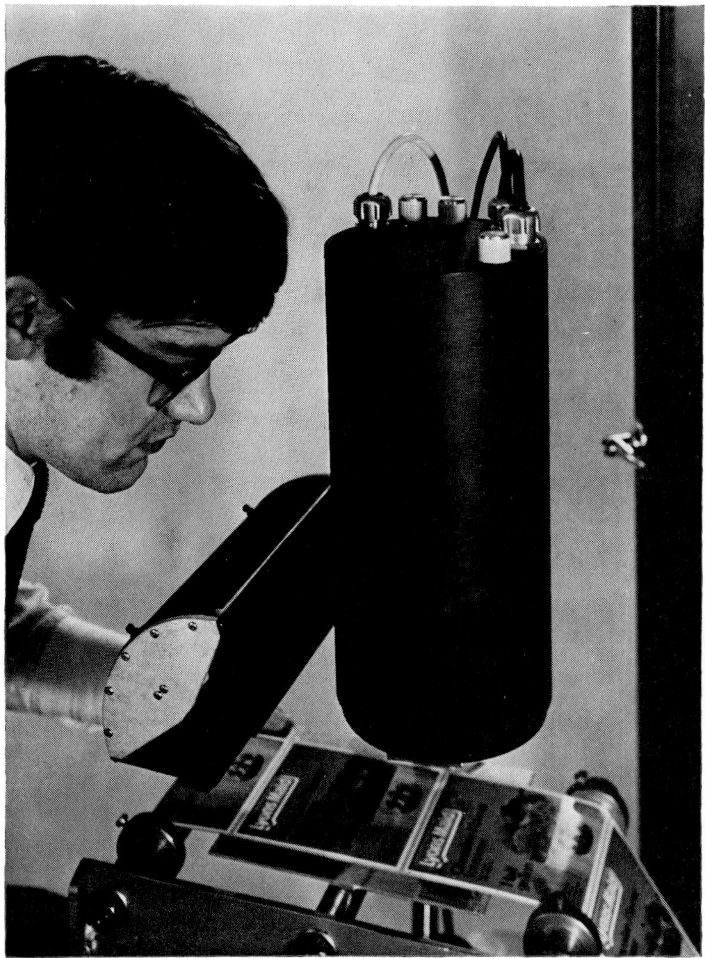

Figure 4.   Print inspection equipment installed on label-reeling machine.

Inspection can be carried out at speeds of up to 20 items/second (representing an on-press speed of up to 200 m/min) and faults which are detected include print-to-print misregister, missing print, spots and smears, and colour variations. The instrument output at present is an alarm signal indicating the presence of a defective product.

The instrument performance depends to a large extent on the patterns under inspection, and especially on the degree of variation which occurs among acceptable products.   As a general guide, misregister of less than 1 mm can be detected on 300 mm wide products and in some cases misregister of 0·3 mm can be identified. Colour differences can be detected in solid or half-tone print and the sensitivity is such that it appears to be possible to detect colour differences at least as small as those normally of interest to the printer.

3. Component inspection

The problem of inspecting components for surface defects such as scratches and handling marks is more difficult than strip products, as previously discussed, owing to the fact that a good proportion of them have curved surfaces. In the optical industry, for example, there is a requirement for the inspection of lenses and mirrors for scratches and digs which may be only just visible to the eye under optimum viewing conditions, but for which the optical component must nevertheless be rejected or reworked. Evaluation of such faults is at present subjective, and there is a requirement for a means of making a quick, objective assessment of surface defects, thus eliminating the need for visual examination by skilled inspectors.

A gauge which is capable of detecting scratches, digs and poor polish on lenses is now undergoing trials. A simplified diagram of the optical system is shown in figure 5. A laser beam is imaged by a microscope objective and two

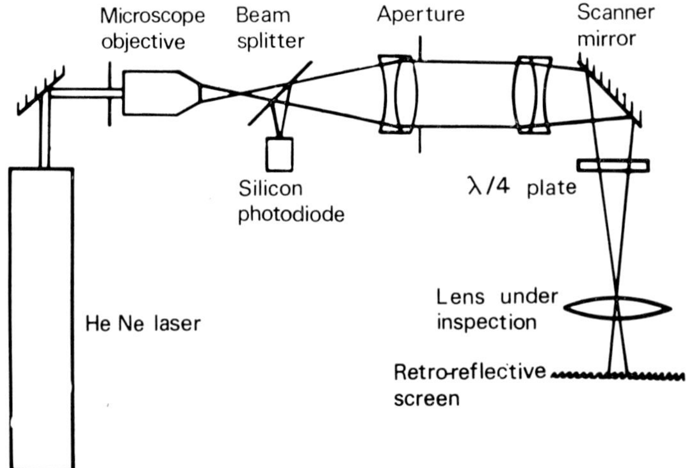

Figure 5. Optical system for inspection of surface defects on lenses.

doublet lenses, to form a fine spot on the lens under inspection. Since the scratches to be detected may be very small—in the range 2–20 μm wide—a small light spot must be used. In order to concentrate a realistic amount of energy into this spot, a 3 mW helium-neon laser is used as the light source. The spot is scanned across the surface under inspection by means of a vibrating mirror.

Behind the lens is a screen of retro-reflective material, which has the property that it returns the beam along its direction of incidence. Hence, for any power of lens under inspection, the scanning light beam is always returned along its own path, back through the optical system to the photodiode. If any light is scattered out of the beam owing to the presence of a scratch or other surface defect, this light is deviated from its original direction.

After retro-reflection, it is prevented from re-entering the optical system, and the output from the photodiode shows a sudden drop, which indicates the presence of a scratch on the lens. The electronic processing circuitry detects any drop in photodiode output which exceeds a pre-set threshold level. Thus the sensitivity of the system can be varied to suit a wide range of requirements.

To inspect for scratches over a curved lens surface, a special scanning arrangement has been devised which maintains the light spot in focus over the entire surface. The spot is made to move in a spiral fashion, starting at the centre of the lens and moving outwards in a circle of slowly increasing diameter to cover the entire surface. During this time, the objective lens is moved by a motor in such a manner that the focus of the beam follows the curvature of the surface. By this means, the spot may be kept in sharp focus over lenses of 100 mm diameter with minimum radius of curvature of 100 mm. The movement of the focusing lens is controlled electronically, the curvature to be followed being set on a dial on the processor unit.

A photograph of the prototype gauge is shown in figure 6. The instrument output is in the form of an oscilloscope display showing a plan view of the lens

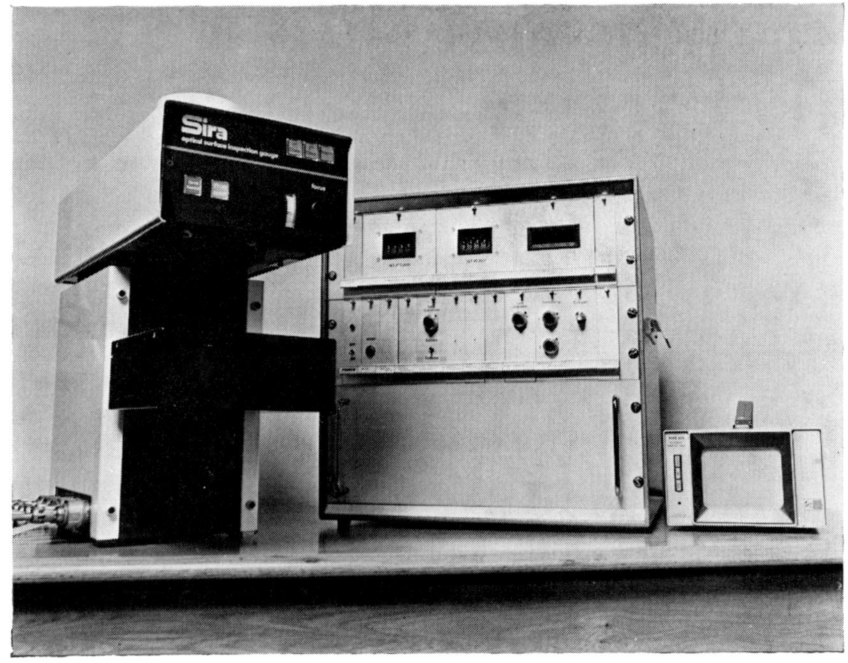

Figure 6.   Optical surface inspection gauge.

with scratches exceeding the required severity indicated by bright lines. Also, the defect pulses are counted to produce a digital value related to the total area of scratches on the surface, the total pulse-count being displayed on the electronic unit. This value is compared with pre-set values and the lens is automatically classified as first quality, second quality or reject, the classification being shown on indicator lamps on the optical scanning unit.

Tests on the prototype gauge indicate that it is possible to detect even very small scratches and sleeks which are hardly visible to the naked eye. The system is also able to detect tiny dust particles, and is capable of assessing surface cleanliness as well as surface damage. New applications of the gauge to study the micro-finish on such products as semiconductors and crystals are currently being undertaken.

4.   Inspection of blade edges

So far we have dealt with the inspection of two- and three-dimensional surfaces but the one-dimensional problem also exists. Large numbers of metal blades are produced annually for use in industrial cutting machines, and rubber and plastic 'doctor' or wiper blades are widely used in the printing and other industries. It is important to ensure that the edges of these blades are of a very high and uniform quality if they are to perform satisfactorily. Inspection is carried out by trained inspectors who examine edges with the aid of profile projectors to check dimensions and to detect edge faults such as nicks and protrusions. Such inspection is laborious and fatiguing, and it frequently happens that defects are missed and substandard products allowed to pass.

An instrument has recently been developed to overcome the problems of visual inspection. It allows blades to be inspected automatically for edge faults, taper and other dimensional variations. It operates on a light obscuration principle by monitoring the change in intensity of a small light beam projected across the edge under inspection as the blade moves past the inspection head. Small defects such as nicks cause localized changes in the electronic signal generated by the photodetector and by suitable signal processing these impulses can be separated from the much slower variations arising from gradual changes in product dimensions such as taper or slow undulations. The instrument output is in the form of go/no-go indicator lamps and buzzer, the sensitivity being adjustable to ensure that blades having defects or dimensional changes larger than pre-set values are rejected. An oscilloscope display of the processed signal can be used to enable the exact location of small faults to be identified.

A prototype gauge has undergone tests in the laboratory and it has proved possible to pick up edge nicks as small as 12 μm deep and also to monitor slower variations in blade profile. The instrument inspects 25 cm blades along their entire length in around 1 sec. Further development is envisaged, aimed at improving the sensitivity to enable edge defects down to 2 μm in size to be detected: this will necessitate the use of a system to monitor precisely the blade edge position before inspection for defects.

5.   Conclusion

The advanced and specialized knowledge required for the design and development of modern electro-optical instruments is encouraging an increasing number of manufacturers and users of such systems to seek the help of instrument-orientated contract research and development organizations. In this way firms too small to support their own research and development teams can compete successfully with larger companies. The area of automatic inspection, where applications vary from firm to firm, is a good example of where special purpose electro-optical equipment is required and has already been successfully applied.

The ophthalmic industry structure and the application of mechanical handling to the production of pairs of spectacles in Great Britain

H. J. BOOTH

Autoflow Engineering Ltd, Lawford Road, Rugby, England

Abstract. After a general description of the UK ophthalmic industry, this paper describes a continuous flow factory producing 4000 pairs of spectacles per day and run to hourly production targets.

1. Introduction

The ophthalmic industry across the world is coming into a period of significant change in relation to:

1. The changing structure of the industry through the merger and integration of companies.

2. The machinery required by the optical industry as a result of engineering developments and economic pressure for lower unit-cost production.

3. The application of conveyerization to the mass production of one-off pairs of spectacles in a prescription factory.

This article divides into two parts:

1. The first part considers briefly the UK ophthalmic industry structure to enable the readers to compare the different operating pattern to that in their own country.

2. The second part considers the application of continuously moving conveyers to the production of completed pairs of spectacles.

2. UK optical industry

Some general facts relating to the optical industry in Great Britain are as follows:

Number of people who wear spectacles	25 million +
Number of registered ophthalmic opticians	6100
Number of registered dispensing opticians	1400
Gross expenditure in the ophthalmic sector	£100 million
Number of all ophthalmic optical establishments	4500 +
Number of dispensing optical establishments	630 +
Average interval between sight tests (all)	3·3 years
Annual volume of sight tests	7·5 million
Total pairs of lenses dispensed	9·2 million
Ratio of people : refractionists	7800:1

Over the last 10 years, there has been significant re-thinking in the optical industry as a result of changes that have taken place throughout British industry during the period when many new industrial groups were formed through mergers. The pattern of the ophthalmic industry in Great Britain is basically a five-tier structure as shown in figure 1. The structure is such that Chance-Pilkington Ltd is the main supplier of lens blanks to the industry, the bulk of which feed the mass-production companies of finished and semi-finished lenses.

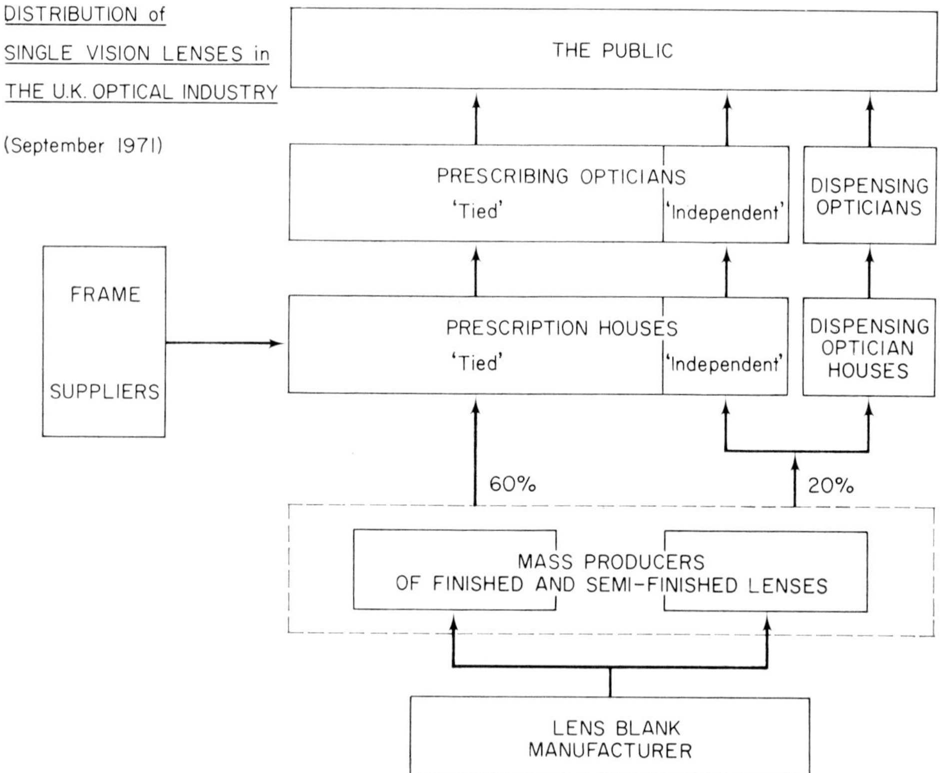

Figure 1. Distribution of single-vision lenses in the UK ophthalmic industry.

In Britain, there are two main producers:

1. UK Optical Holdings Co Ltd incorporating the M. Wiseman Co.

2. British American Optical Ltd.

These two companies, in turn, supply the finished and semi-finished lenses to the prescription houses (figure 2). The prescription houses are independent companies providing a service to the prescribing optician, and there are about 200 in total. The prescribing optician and the dispensing optician form the retail outlet from the prescription houses to the customers in the High Street, and there are about 5000 in total.

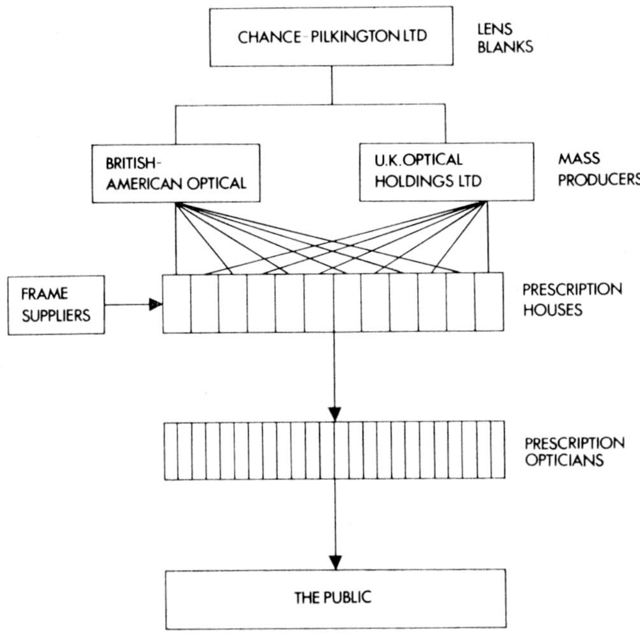

Figure 2.   The role of Chance-Pilkington Ltd.

Unlike the system elsewhere, the prescription house/laboratory undertakes the full service of providing the complete pair of spectacles; that is, each laboratory would have a stock of frames, together with generating, smoothing and polishing capacity, plus lens edging equipment, and would complete the whole spectacle prescribed for the customer by the retail optician.   The professionally qualified optician or ophthalmologist only conducts a sight test at the retail shop or practice.   Some retail outlets are 'tied' to a prescription house/laboratory as in the case of, say, the Dollond & Aitchison Group of retail shops but, again, the majority of prescription houses are too small and tend to provide only a localized town or city service.

The optical industry is today still very fragmented and has only started to re-structure and optimize its production and earnings capability.   In respect of the retail side of the UK industry, the Dollond & Aitchison Group has developed into the world's largest fully mechanized central prescription factory, as a result of the merger of three separate operations:

1. The Harrison Group + 68 retail outlets.

2. Hudson Verity Group + 110 retail outlets.

3. Dollond & Aitchison + 108 retail outlets.

The outcome of this merger, which could be justified on the basis of establishing significant cost reduction through integrated planning and the implementation of a centralized production philosophy, has been:

1. A prescription factory employing 500 people in total and producing about 4000 pairs of spectacles per day with glass lenses, each to an individual prescription order.

2. A separate prescription factory producing 700 pairs of spectacles per day with plastic lenses.

3. A network of over 300 retail outlets integrated with the prescription factory through close management control.

4. A major source of some 300 professional ophthalmic opticians.

5. Dollond International Ltd, which is a group company looking after the long-term interests of the separate overseas company purchases, which to date are located in Europe.

However, with the economic pressures for lower-cost production, more UK optical companies will form new associations within the industry over the next two or three years.

3. The production of spectacles

Let us consider the production of spectacles and the philosophy that can be applied in achieving various levels of output in the prescription factory. The production of spectacles is basically the production of one-off jobs. A recent definition states that ophthalmic prescription work consists of the one-off production of a large number of different items to an almost infinitely variable but precise specification. To achieve this has, in the past, required the total reliance on skilled operators working at each stage of production:

1. Generating toric and spherical curves.

2. Smoothing and polishing.

The majority of prescription laboratories fall into the 100–400 jobs per day range and the order handling and work flow is generally poor and untidy. The progressing and general efficiency is made more difficult by the continuous handling of the work trays. Looking at the method of production critically in many prescription houses, the most noticeable feature is the 'people movement': people walking about carrying piles of trays; other people waiting for trays to be brought to them; trays with different colours and with dates and other various identifications. The situation is manageable at 100–300 jobs per day but at 3000 jobs per day the production problems become extremely difficult, if not impossible.

The growth in the output from prescription houses has demanded an improvement in performance and the application of advanced technology to machinery designs. For example, Autoflow Engineering Ltd has a range of prescription toric generators which provide a major change in production capability (figure 3).

In considering the development of toric generators, there are a number of aspects of the design of the lens generator that have changed significantly over two decades (table 1).

Figure 3. Fully automatic toric lens generator complete with punched tape input: the Autoflow Supermatic.

Table 1. Development of toric generators.

	1954	1972
Range of curves	Base ± 2·00/10·00D Cylinder ± 3·00/15·00D	± 2·00/9·90D ± 3·00/14·00D
Diameter of diamond wheel	71 mm	86 mm
Grinding time	3 min	15 sec
Size of lens	60 mm	75 mm
Control	Manual + hydraulic	Solid-state circuitry components and digital/ analogue conversion system
Operational aspects	Manual and skilled operator	Fully automatic and unskilled operator
Output per hour	22 surfaces	100 surfaces

The development of high output machinery means that there must also be parallel improvements in the interface materials.

1. Diamond wheels must be capable of a much faster rate of stock removal and yet offer as long a life as possible. An average life of a 15 concentration Grade 60/72 wheel is 8000–10 000 surfaces.

2. Polishing powders must be capable of polishing a greater number of surfaces per kilogram. An average life is 500–600 surfaces/kg. A lot depends on how frequently the machinery is cleaned and new powder used.

3. Zinc foils and felt or polyurethane pads all help to reduce the smoothing and polishing times, providing the grinding accuracy is high. A prescription machine can give an average time as follows:

   Smoothing time   ...   4–5 min

   Polishing time   ...   6–7 min

It is this type of machine and material development which demands a new look at work-flow methods, in order to fully utilize or optimize the output capability of the machinery. Keeping in mind the trends in machinery and possible application in a mass production type of environment, Dollond & Aitchison, as a company, forced a whole new production philosophy to be evolved.

The Dollond & Aitchison retail pattern of distribution can be seen in figure 4, resulting from the merger of the three UK prescription companies. Instead of seven separate prescription laboratories, the Dollond & Aitchison Group have established a central glass lens prescription factory which today produces 4000 pairs of spectacles per day.

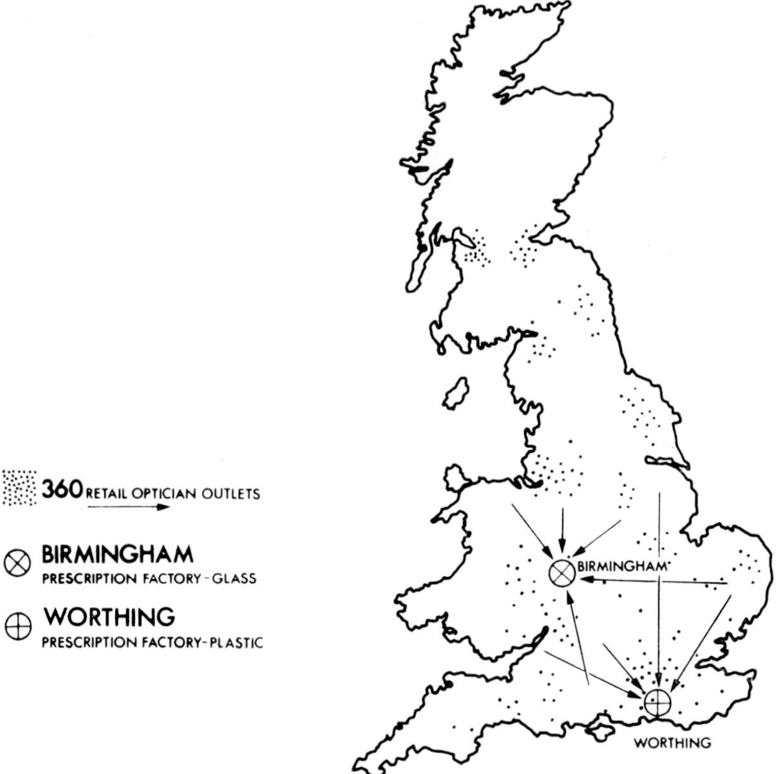

Figure 4.   Distribution structure of the Dolland and Aitchison group.

In addition to the central factory at Birmingham, the Dollond & Aitchison Group established a central prescription factory on the South coast producing about 700 pairs of spectacles per day with plastic lenses.   To justify the full benefit of building a central prescription factory with over 300 retail outlets, a number of criteria had to be established:

1. Was a centralized factory production possible and practical?

2. Was the general postal service in Great Britain reliable to ensure a 24/48 h turnround of work?

3. Was it possible to move the work input on a mass production basis, in view of the fact that each completed product in any day would be different?

The solution to the problem lay partially in an analysis of the differences between prescriptions and not wholly in the method of handling.   However, a critical consideration was the continuous flow of work between sections.   In conjunction with a British mechanical handling company, a total conveyer system was laid down at the central production factory in Birmingham.   Figure 5 shows diagrammatically the flow of work between the various sections and the layout of the main motorized conveyers and gravity roller conveyers.

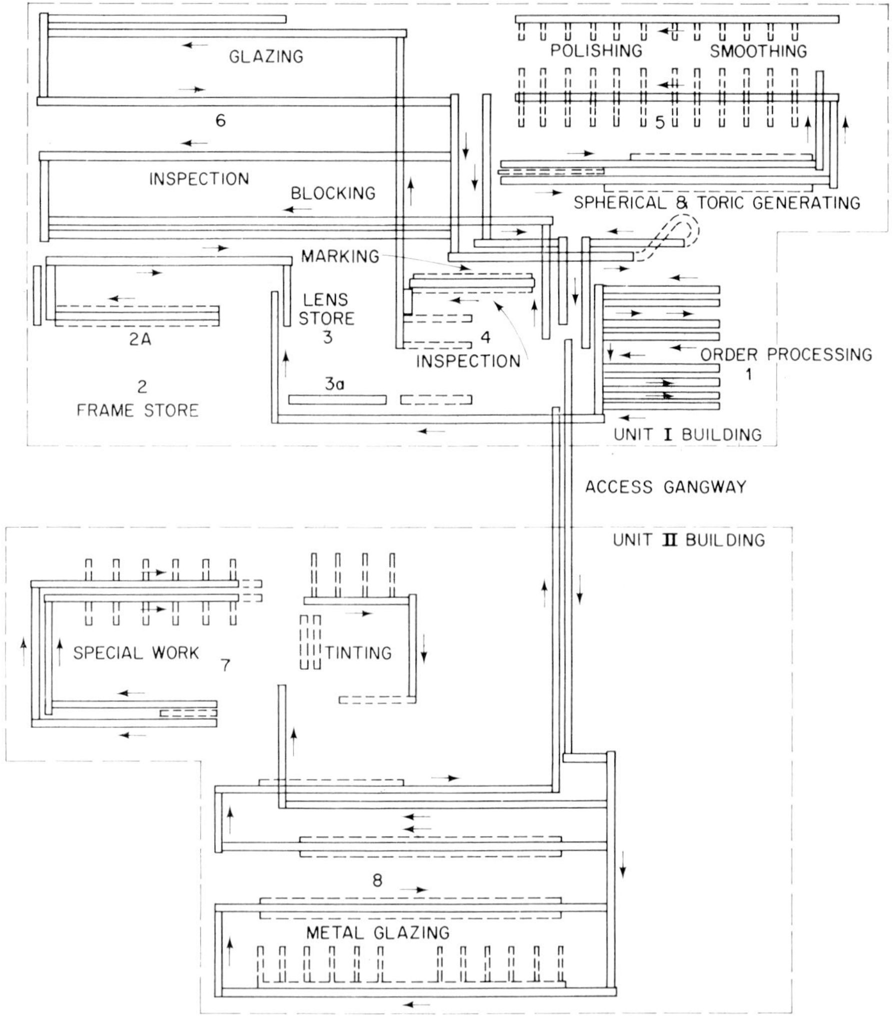

Figure 5. Flow of work around Dolland and Aitchison factory in Birmingham, England.

4. Post room and order processing

Twenty receiving stations accept incoming work. Each station deals with one group of practices (opticians and staff located in the retail shops) within the company. At the end of the day, finished work returns to the same stations for despatch. After the incoming work has been recorded, the orders are sent to the frame store, where some 260 different styles of frame are held in stock. The frames are kept on racks arranged alphabetically according to type and hung on spigots under the frame name shown in the catalogue, which is available at each retail shop for customer selection. By hanging the frames vertically under the colour description and horizontally by measurement, the Stores staff can select any frame within an average of 10–15 sec. The frame, plus the order form, is placed in a tray and sent to the plug gauge former collection board (figure 6). In a flow system, using labour with less specific skills than in the

Figure 6.   Frame stores and plug gauge former collection.

past, it is important that no one should be called upon to interpret and take action on an order beyond their level of understanding.   Therefore, the factory is run on a go/no-go system, under which anything that will go at any stage passes on in the least possible time.   No-go items are side tracked to sections where they can be dealt with by people or machines suitable for the requirements of the order.

The plug gauge former has three important functions in a flow system:

1. It is a 1/1 glazing template which allows us to lock the automatic edging machines and avoids the operators having to size the machine for each order.

2. It is a plug gauge for the eye shape and size which is offered to the frame, proving that the lens will fit the frame on a machine that has been maintained to cut that size plus allowance for the bevel.

3. It acts as a size and shape check on fresh supplies of frames from the manufacturer.

Special formers are cut to suit non-stock frames.

Once the tray is ready with its spectacle frame, plug gauge for glazing and order form incorporating the prescription, verification will be completed by the computation section. The verifying procedure will check the prescription to decide if a stock lens can be selected from the stores and allow the tray to be sent directly to the glazing section.

In fact, 60% of all work passes directly into the glazing section.

Figure 7. Overall view of the surfacing and glazing department, producing 4000 pairs of spectacles per day.

However, when the verification shows that the lens cannot be fulfilled from stock, the computation section undertakes the transportation of the prescription for surfacing and recording:

1. Type of lens blank or semi-finished lens.
2. Curves required.
3. Centre thickness.
4. Edge thickness.
5. Selection of lap tool number.

With the former, the trays are taken by conveyer to the lens store, where they receive either a stock finished lens or a lens blank or semi-finished lens. From the lens store the tray is sent to the lens acceptance area, where the lenses are marked for axis and subsequently sent to glazing or surfacing.

Figure 8.   Toric smoothing machines with work arriving on lower conveyer and being taken away on upper conveyer in surfacing department.

5. Surfacing

Of the 4000 complete jobs per day, 60% are satisfied from stock lenses. Therefore, 1600 jobs, or 3200 lenses, require to be surfaced (figure 7).   Jobs directed to the surfacing department are conveyed to the metal-blocking section and on to the spherical or toric generators. The conveyer system comprises powered drives pulling 6 in. wide belt conveyers and transporting the 5 in $\times$ 7$\frac{1}{2}$ in $\times$ 2 in high trays to the sphere and toric smoothing and polishing sections. At each smoothing and polishing section there is a number of operators, each controlling three machines. A gravity-roller conveyer feeds the work to each operator by the action of an automatic pneumatic sweep gate which diverts the work travelling on the main belt to the sections (figure 8). Surfaced lenses are then de-blocked in boiling water and returned for inspection and marking for axis and power checks before glazing.

Figure 9.   Conveyers feeding the frame heaters and de-starring section in the glazing department.

6. Glazing

The three main conveyers are used to transport the work to the operatives in the glazing section (figure 9) where:

1. Lenses are edged on automatic machines grinding the lens to size in accordance with the former.

2. Springing-in. Automatic time cycle heaters are used, basically ensuring increased output by using the operator's time for the actual springing-in of the lens into the plastic frame and not for pre-heating.

Figure 10.   Final inspection for lens axis and power.

7.   Inspection

The final inspection (figure 10) is a three-part operation:

1. *Visual inspection.* This is an inspection of the frames and lenses to a comprehensive list of visual checks.

2. *Focimeter inspection.* Here the major checks for power axis and centres are made. Any failures are sent to a recovery section for final assessment.

3. *Final look.* To ensure that all the instructions on the order have been fulfilled.

The customer's pair of completed spectacles is now passed by conveyer to the despatch and packing section, which is also the same receiving point. Special work involving high-power lenses, lenticulars, tinting and metal glazing is sent to the Unit II area shown in figure 3. All trays travelling through the various sections are monitored by photoelectric cells housed at strategic points in the flow system. Repeater counters are located in the works director's office and each departmental manager has repeater counters showing the output of his department. All sections are working to hourly targets. Because of the development of a production philosophy that is designed for continuous flow, it immediately imposes a tremendous pressure on the administration of the central factory to ensure that there are no hold-ups, such as:

1. Out of stock lens sizes.

2. Out of stock frames.

3. Lack of consumables such as polishing powders or toric tools.

4. Lack of daily and weekly maintenance checks.

However, the purpose of this article is to show that the optical industry is changing and the role of the machinery maker is becoming more dynamic and influential. However, the real change and progress in the future of optics will only occur if there is much more open discussion on the techniques of manufacture and the solution to technological problems.

Optics is changing, but the best has yet to come.

Thermal imaging lenses

P. L. WRIGHT

Rank Optics Taylor Hobson, P.O. Box 64, Stoughton St., Leicester, England.

Abstract. This paper describes some lenses for use in the 8–14 μm waveband now available from our company and gives an indication of the levels of performance which can be achieved.

In recent years our expertise in the manufacture of non-spherical refracting and reflecting surfaces has played an important role in the research and development of infra-red lenses under British Ministry contracts. Such surfaces have considerable significance for the lens specifications that are required and for the infra-red transmissive materials which have to be used in conjunction with the new pyroelectric vidicon detectors and other similar devices.

1. Irtal 1 and Irtal 2

Figure 1 lists a number of representative examples in the range of lenses giving their specification in terms of focal length and relative aperture. The range covers relative apertures from f 1·5 to f 0·7 and it will be appreciated that the longer focal length lenses are quite large, having diameters of about 200 mm. These lenses have been given the trade name Irtal.

Figure 2 shows the optical construction of Irtal 1, as being a very simple single element made from germanium, which is an extremely useful material for use in the 8–14 μm band because of its good transmission, its high refractive index and its low dispersion. Its dispersion, i.e. the variation of refractive index with wavelength, is small enough to avoid the necessity of having compound components using two different materials to correct chromatic aberrations.

In this simple construction two design variables are available: one is shape, which controls the coma aberration, and the other is the equation defining the shape of a non-spherical surface, which controls spherical aberration. Consequently an aplanatic correction is possible and by expanding the aspheric equation to include the tenth power of the aperture, all high-order spherical aberration can be removed to yield a diffraction-limited performance in the region of an axial image point.

Choice of surface to be aspherized is not critical to the design but for practical convenience we make the second surface aspheric and leave the external front surface as a normal spherical surface.

Of course, the high refractive index of germanium means that anti-reflection coating of the surfaces is essential. If the surfaces were not coated, transmission efficiencies of less than 40% would be the best obtainable through the two surfaces of Irtal 1. The required coatings differ from those adopted in visible-spectrum optics because of the high refractive index of the substrate and the much longer wavelengths involved. Special coatings are available which yield transmission efficiencies in excess of 95% per surface, so that overall efficiency

 P. L. Wright

IRTAL RANGE

Irtal 1	100 mm f 1.0	Single Element
Irtal 2	100 mm f 1.0	Double Element
Irtal 3	100 mm f 0.7	Double Element
Irtal 4	50 mm f 0.7	Double Element
Irtal 5	30 mm f 1.0	Double Element
Irtal 7	200 mm f 1.0	Double Element
Irtal 8	300 mm f 1.5	Double Element

Figure 1

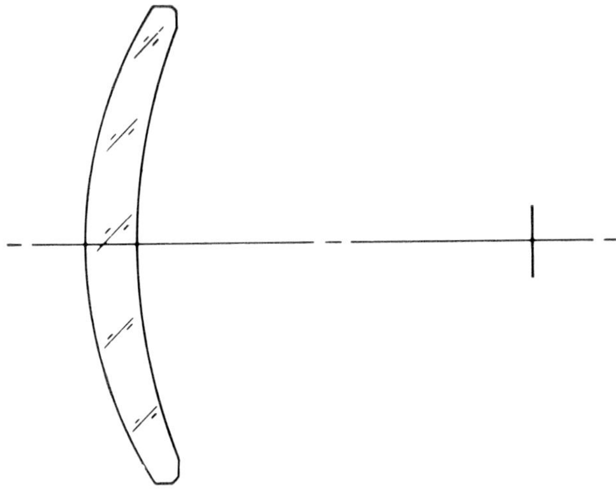

Figure 2

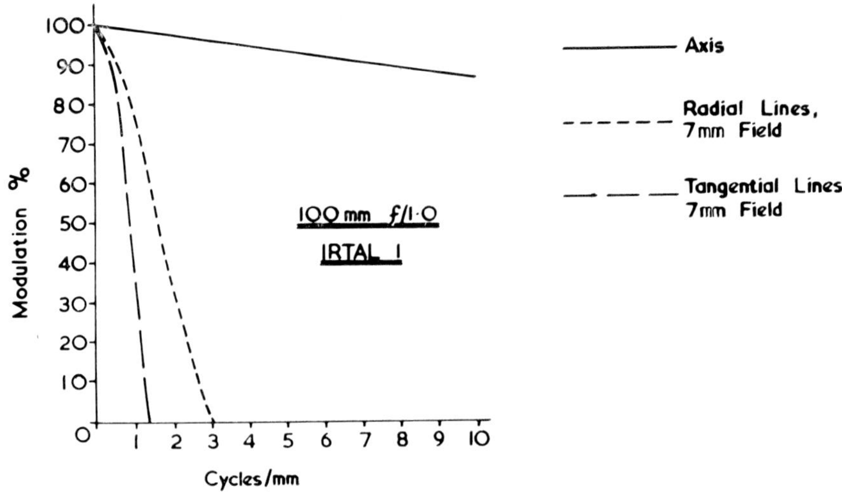

Figure 3

is more predominantly dependent on the quality and volume of the germanium used in a particular product.

Figure 3 shows the monochromatic m.t.f. performance of Irtal 1 and includes the effects of diffraction. The curve for response on-axis corresponds to the theorectical diffraction limit for a lens of this relative aperture, showing that all orders of spherical aberration have been eliminated. Because the construction is simple, off-axis performance is severely limited by the presence of astigmatic errors, as indicated by the response curves at a field radius of 7 mm.

Although defocusing can improve the off-axis response at the expense of the axial response, this simple construction is not recommended in circumstances where useful fields of view are required.

More design variables are required to bring off-axis aberrations under control and figure 4 shows one development adopted in Irtal 2, in which an additional lens element is positioned immediately in front of the focal plane.

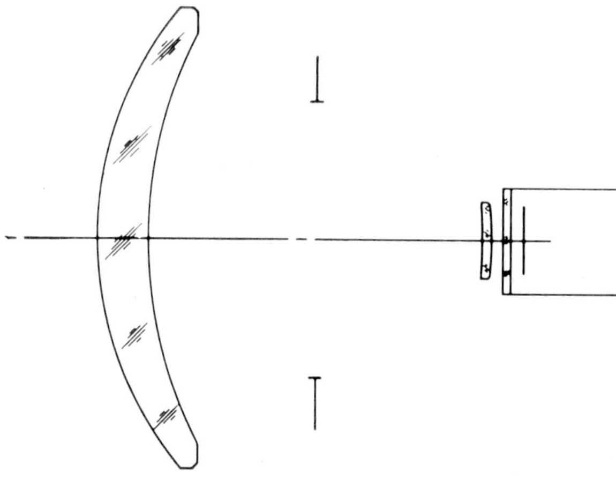

Figure 4

The main function of the second element is to control the astigmatic errors rather than Petzval-field curvature, so the element does not have to have significant refractive power. Consequently the first element has similar characteristics to that of Irtal 1 and there is no impairment of the on-axis performance.

To obtain the required control of astigmatism, the second element necessarily has one of its surfaces aspherized.

The use of two non-spherical surfaces in constructions like this is fundamental to the design concept and should not be considered as a refinement applied to a system of spherical surfaces.

The problems of reproducibility of non-spherical surfaces in a multiaspheric system are more stringent and somewhat different from those encountered when only one aspheric is required. This Irtal 2 lens has demonstrated that the use of highly specialized tape-controlled machinery for surface generation and adequate equipment for performance assessment can provide the required precision under production conditions. Its success in this sense has encouraged the development of other multiaspheric lens systems to be described later in this paper.

Figure 5 shows the m.t.f. response of Irtal 2. In this case the response shown is polychromatic over the 8–14 μm band in accordance with the spectral response of the complete viewing system including spectral atmospheric windows.

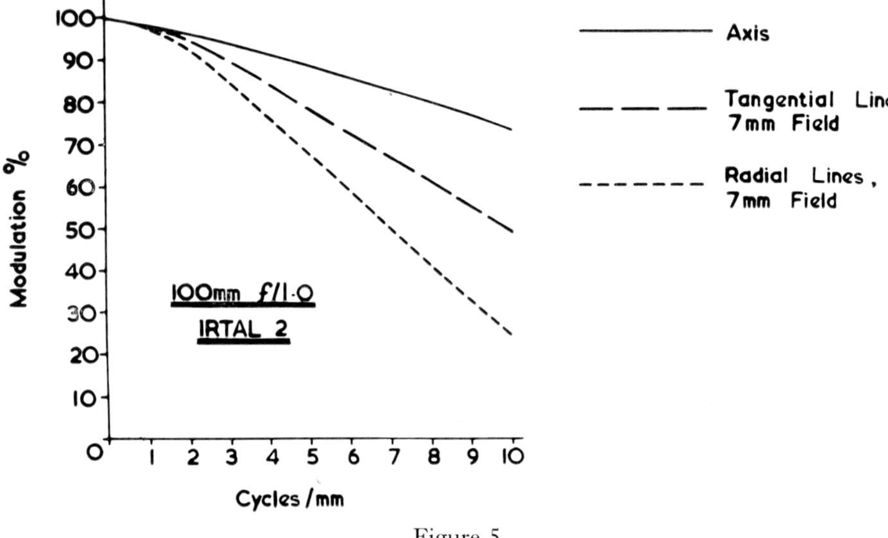

Figure 5

The axial performance is identical to that of Irtal 1, whilst the off-axis performance is greatly improved. Further improvement of the off-axis performance in this type of construction is limited by the introduction of small residual aperture aberrations associated with the correction of astigmatic errors.

Following the success of Irtal 2, the optical designer has taken full advantage of the freedom to utilize more than one non-spherical surface as a means of obtaining even better control of all aberrations and other basic types of construction have been introduced.

2. Irtal 3

Irtal 3 was the first of these new types. Figure 6 gives an impression of it mounted on a thermal imaging camera. Its optical construction can be seen to be somewhat analogous to a conventional Petzval lens (figure 7). The refractive power of the lens has been distributed more evenly between the two elements so that the aspheric surface of the second element becomes more effective whilst the shape of the second element provides an extra design variable.

Irtal 3 has a number of advantages compared with the earlier Irtal 2 design.

1. The relative aperture is increased from f 1·0 to f 0·7.

2. Its off-axis performance is more uniform over a wider field.

3. It holds its performance level over a wider range of object distances.

4. It has a much longer back focal distance, permitting a wider choice of image detectors.

5. It requires the use of only about half the volume of germanium compared with other lenses of similar specification.

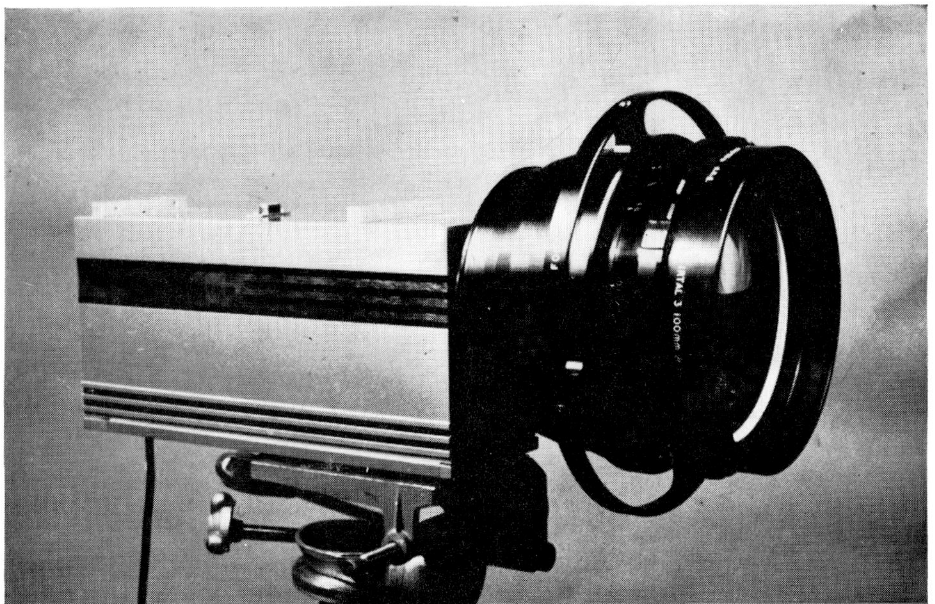

Figure 6

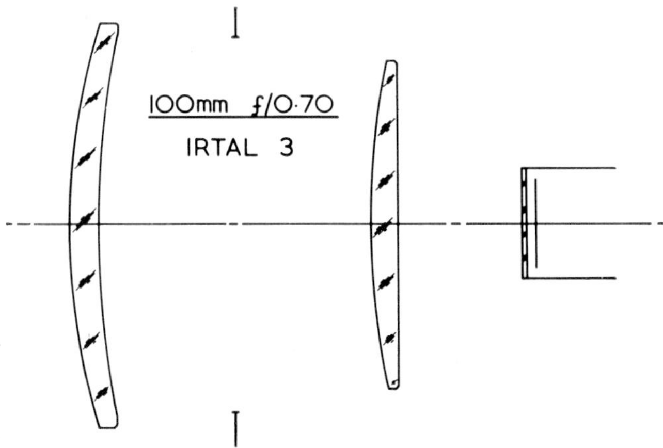

Figure 7

Figure 8 shows the performance level that can be obtained with Irtal 3.   At a spatial frequency of 5 c/mm, the high modulation levels are uniform across the Vidicon image format in the presence of an aperture increased from f 1·0 to f 0·7.   Acceptable performance extends over a larger image format of about 24 mm.   Because the field coverage of Irtal 3 has been increased in this way, it becomes possible to scale the lens down to a 50 mm focal length to provide wider angular fields of view when it is used on the standard 16 mm diagonal image format.

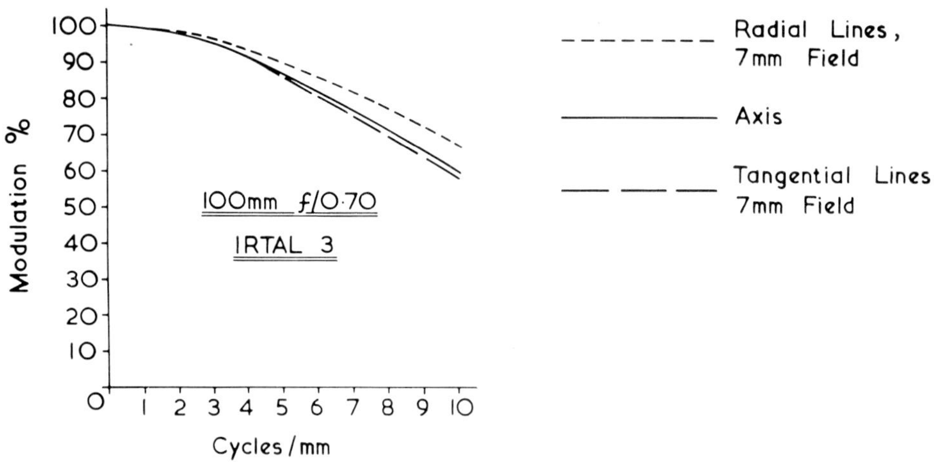

Figure 8

3. Irtal 4

Figure 9 shows 50 mm f 0·7 Irtal 4 mounted on a camera and figure 10 shows the optical construction to be the same as Irtal 3. Modulation levels at a spatial frequency of 5 c/mm are quite acceptable over the 16 mm image format (figure 11).

4. Irtal 5

Some applications demand angular fields of view wider than the 20° field provided by Irtal 4 and for these purposes it is necessary to search for other basic types of construction. One solution is to adopt an analogue of the double

Figure 9

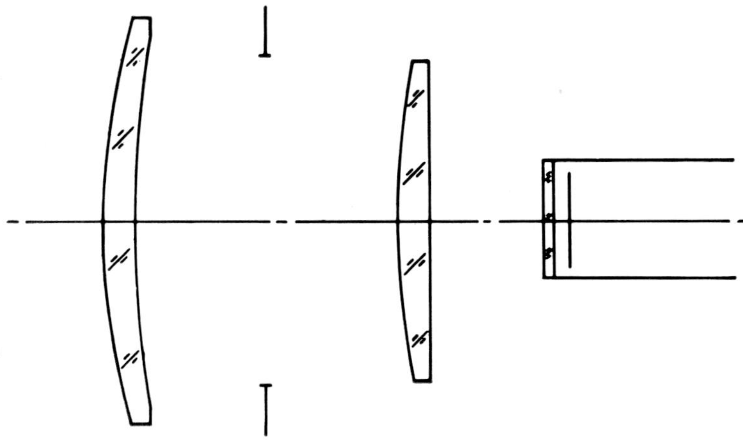

Figure 10

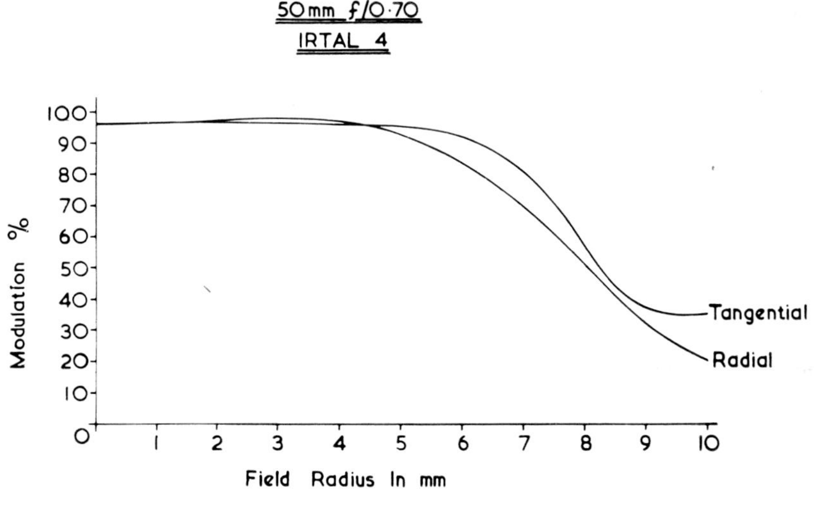

Figure 11

Gauss type of construction. An example of this type is shown mounted on a camera (figure 12). Its specification is relative aperture f 1·0 and focal length 30 mm. Used with the standard 16 mm image format this corresponds to an angular field of view of 30°. The double Gauss family likeness can be seen in figure 13. A powerful computer design program has optimized the power contribution between the two elements in much the same way that it did in the Petzval type of construction and has maintained the maximum effectiveness of the two aspheric surfaces located in each of the elements.

In a wide angle lens of this construction there is, inevitably, a different set of compromises in the use of the aspheric surfaces to correct both axial and field aberrations. Consequently the maximum relative aperture of this type of construction is limited to f 1·0. This is, perhaps, a small sacrifice relative to a 50% increase in angular field coverage.

Figure 12

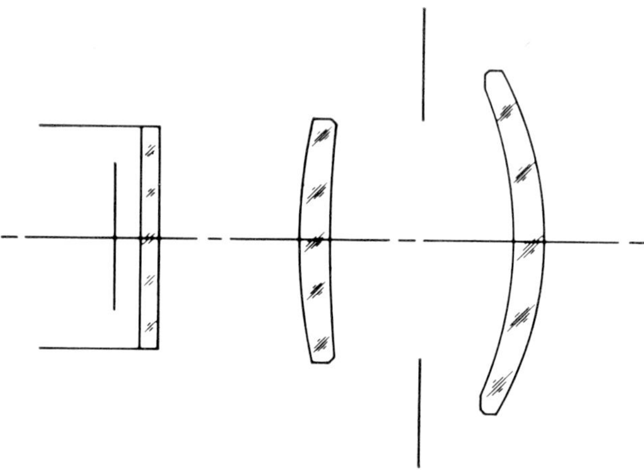

Figure 13

Figure 14 shows the performance of Irtal 5 expressed as modulation at a spatial frequency of 5 c/mm. This wide angle lens is completely free from geometrical distortion errors and achieves good uniformity of thermal energy in the image out to the corners of the image format. Image brightness or energy level in the corners exceeds 75% of that on axis when used at maximum aperture.

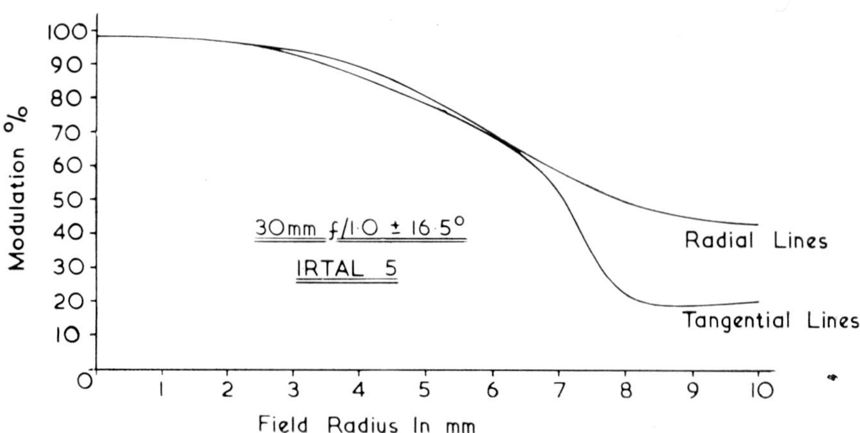

Figure 14

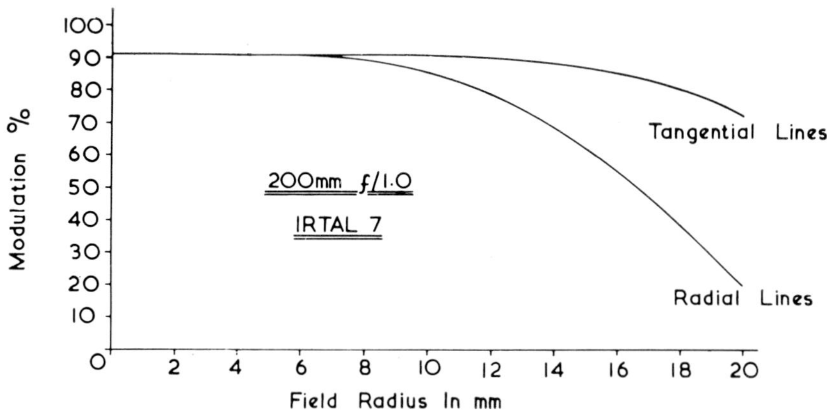

Figure 15

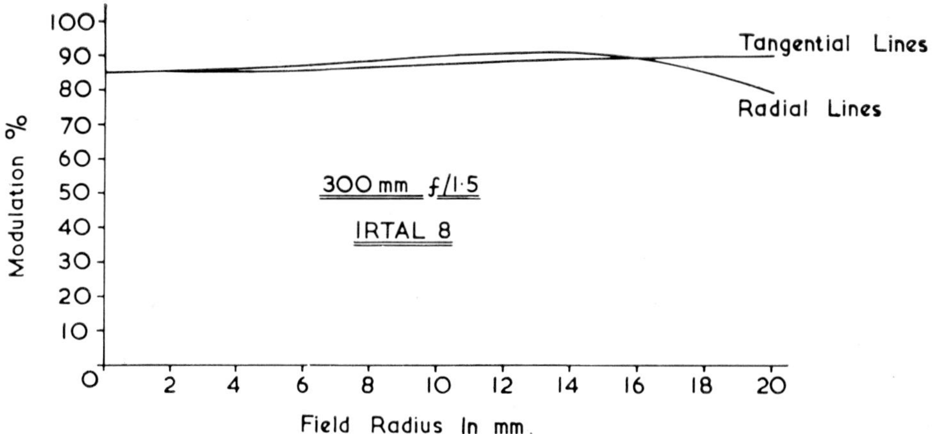

Figure 16

5. Irtal 7 and Irtal 8

Other applications demand large lenses of longer focal length and useful relative apertures. For these we prefer to manufacture large scale versions of the Irtal 3 Petzval-type of construction. Two additional products of this type can be described by way of example.

Irtal 7 is a 200 mm f 1·0 lens whose performance is shown in figure 15. At a spatial frequency of 5 c/mm, modulation levels are good over the 16 mm image format and quite acceptable over larger formats.

Irtal 8 is another version of this type, having a focal length of 300 mm. This lens has a relative aperture limited to f 1·5 for obvious reasons of size, weight and cost. Figure 16 indicates the high level of aberration correction which is necessary in such a large lens.

Recent advances in high speed photography

R. HADLAND

Hadland Photonics Ltd., Newhouse Laboratories, Bovingdon, Herts, UK.

Abstract. This paper describes the recent UK developments in optical and x-ray high speed photography and gives examples of typical industrial uses.

Over the last few years, most advances in high speed photography have resulted in cameras that take streak or framing pictures faster than before. Techniques advanced so rapidly that in 1970, we had a 600 million frame/second camera, but nothing fast enough to look at.

Although we are steadily progressing into the femtosecond region (10^{-15} seconds) for observing such events as relaxation phenomena of excited organic compounds, most of the current developments are aimed at improving image quality and extending the spectral sensitivity into regions other than the visible.

Interest in the sub-nanosecond time region first became significant with the development of short duration laser pulses. Until then, ultra highspeed cameras were of the rotating mirror type, with a one nanosecond time resolution in the streak mode, or image converters with a time resolution of a fraction of a nanosecond. The rotating mirror camera was already fully developed and effort was concentrated on improving image converters, particularly in the streak mode.

Three major problems were encountered:

1. To increase scan speed up to 50-100 mm/nanosecond.
2. To reduce the transit time spread of photoelectrons in the sweep tube to about 2 picoseconds.
3. To intensify the output image by at least 1000 times.

Because of the limited spatial resolution of the camera/intensifier system, which is about 4 line pairs/mm at the recording film, it is necessary to scan at speeds in the order of 50 to 100 mm/nanosecond. This enables the technical time resolution, which is the time taken to scan from one resolvable image to the next, to be brought down to about 4 picoseconds.

Early researchers used laser triggered spark gaps to generate these very fast ramps, but such methods make the camera an integral part of the experiment rather than a separate diagnostic tool. Chains of avalanche transistors have been used with some success by workers in the USA, but they tend to be unreliable in electrically noisy environments. We developed a sweep circuit using ceramic triode valves that could be triggered electrically from a photocell or PIN diode, or optically with a proportion of the laser pulse. With a scan speed of 75 mm/nanosecond, recording time is less than one nanosecond. However, synchronisation to the event is made relatively easy as the circuit only has 100 picoseconds of jitter in its 10 nanosecond delay. This circuit, originally developed for the Imacon 600, is now incorporated with minor changes in the fibre optic output Imacon 675.

The second problem was the limitation of time resolution due to the transit time spreads of photoelectrons. Photoelectrons emitted from the cathode have a spread of emission energies which is usually a large proportion of the energy of the incident radiation. In early tubes, the field strength close to the cathode was low, hence the initial velocity variations caused a wide spread in the arrival times of electrons which had been simultaneously ejected from the cathode. This is because most of the transit time spread occurred close to the cathode where the emitted energies represented a high proportion of the total energy. By placing a mesh extraction grid [1] close to the cathode with a high positive potential on it, the electrons are accelerated rapidly to a high velocity where their initial energies become insignificant.

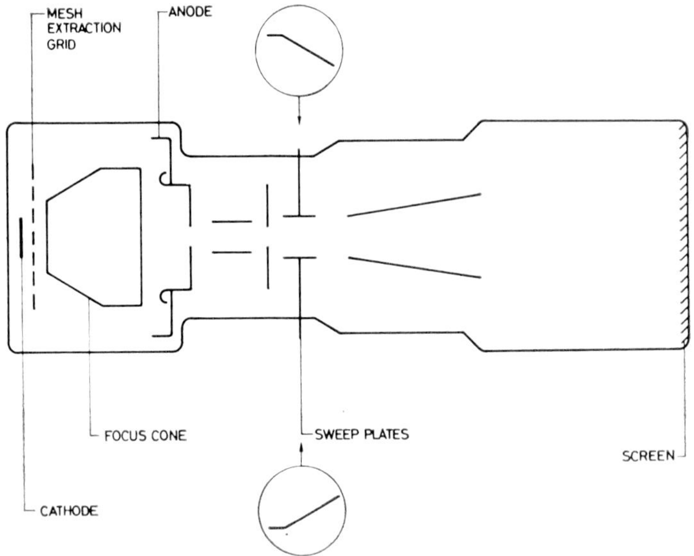

Figure 1.   Essentials of an image tube.

Figure 1 shows the image tube with the mesh extraction grid located 3mm behind the cathode. When a potential of 1 to 2 kV is applied to the grid in relation to the cathode, the initial electron energy, which is usually a maximum of 2 eV, becomes insignificant. This technique is used in the ICC512 camera and Imacon 600 series cameras, and brings the transit time spread factor down to about 2 picoseconds.

The final problem was to record an exposure of such short duration. Until recently most image converter cameras used a plain glass output screen and the image was recorded using a wide aperture lens and fast film. To make a 200 picosecond exposure of a narrow slit image as would be used in streak mode, a beam current of several milliamps is required to make a bright enough image on the screen. At this level, the image starts to distort due to charge repulsion in the beam. For shorter exposures, it therefore becomes necessary to keep the beam current low and increase the brightness of the image using an image intensifier.

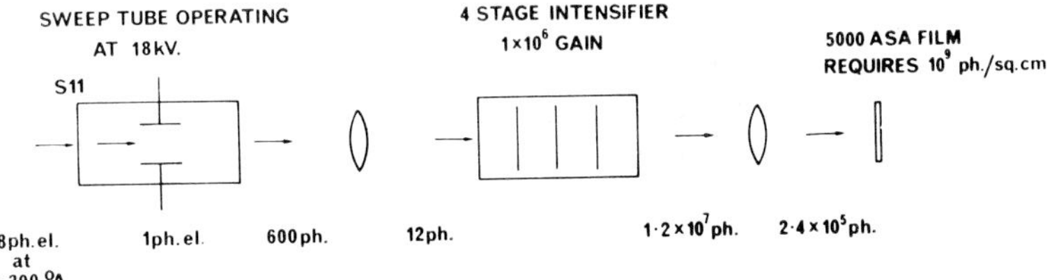

Figure 2.   System with magnetically focused intensifier.

Early systems used magnetically focused intensifiers as these devices offered the best resolution and minimum distortion (figure 2).   As both the sweep tube and the intensifier tube have plain glass screens, it becomes necessary to use lens coupling throughout.   Thus the final relay lens, which has a transfer efficiency of 2%, reduces the maximum gain from 10^6 to 2×10^4 times.   This does, however, give sufficient gain to record the light from single photoelectrons on fast negative film.

Figure 3 shows the Imacon 600 with the latest 4 stage air-cooled intensifier, and the ICC 512 which is of similar design but has certain circuit variations. These cameras both became commercially available in 1971, with a time resolution of 5 picoseconds.   They are still the only commercially available cameras to meet this specification.

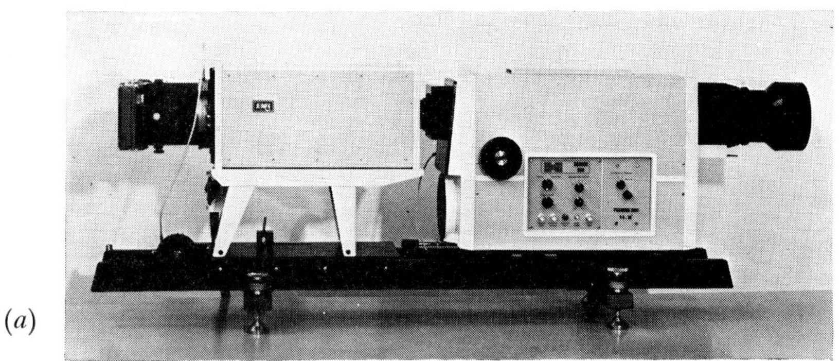

(a)

(b)

Figure 3.

(a) Imacon 600 with 4-stage intensifier;   (b) ICC 512.

I

                    R. Hadland

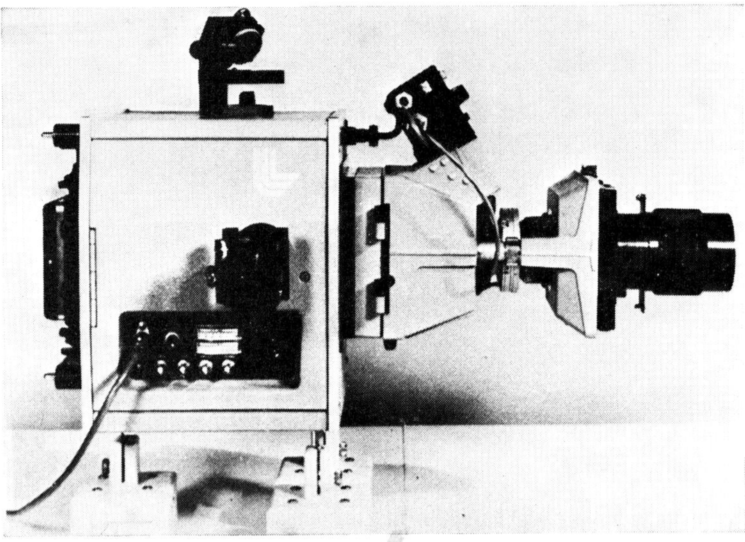

Figure 4.   The Livermore experimental camera incorporating a channel plate intensifier.

When channel plate intensifiers first became available, they appeared to be the obvious replacement for magnetically focused tubes.   A number of experimental cameras incorporating channel plate intensifiers were built in the USA, the most significant being the Livermore camera, figure 4.   This camera incorporates a wafer type channel plate intensifier, in which the imaging into and out of the microchannel plate is proximity focused and does not require a focus cone.   Whilst this type of construction, known as a wafer channel plate is technically the most superior, severe problems in manufacturing make the yield and cost of good tubes unacceptable, and they have not yet been incorporated in commercially available cameras.

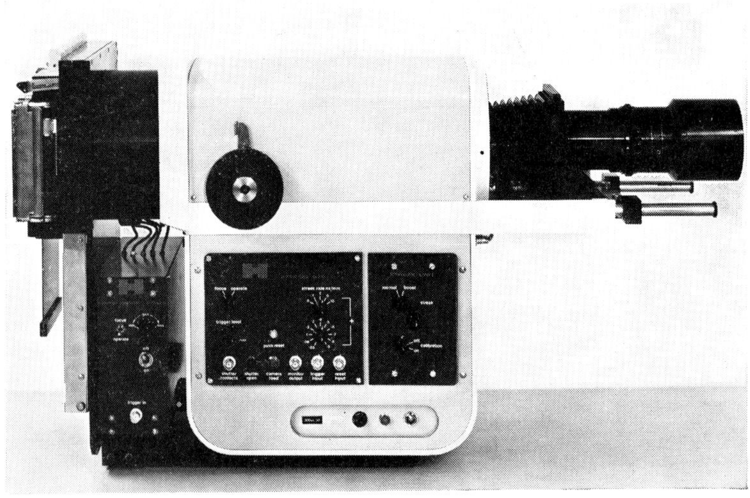

Figure 5.   675 streak camera, commercially available and incorporating a channel plate intensifier.

In the UK, focused type channel plate intensifiers were being developed. These are larger, about 100 mm long as opposed to 20 mm for the wafer type, and also have a slight pin cushion distortion.   However, these tubes are reliable and commercially available, and we have incorporated the Mullard 50/40 type intensifier in the new 675 streak camera, figure 5.   This is the first commercially available streak camera with channel plate intensification.   The Imacon 675 has a fibre optic output screen and the 50/40 tube has fibre optic coupling at both ends.   Thus coupling lenses are not required and the size of the system is considerably reduced.   The recording film is held against the output screen by a foam pressure pad.

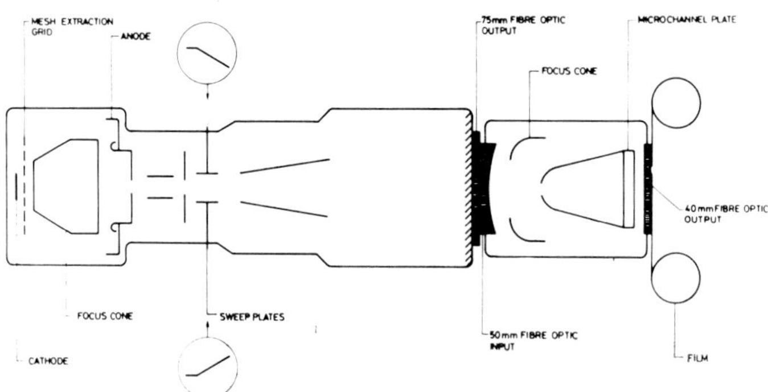

Figure 6.   Layout of 675 and 50/40 system.

Figure 6 shows the layout of the 675 and 50/40 system.   The intensifier records the central 50 mm diameter of the sweep tube and reduces it to 40 mm diameter.   This magnification change occurs in the focused section of the tube before the channel plate.

A comparison between the Imacon 600 and 675 is shown in Table 1.

Table 1.   Comparison of two converter cameras.

	IMACON 600 4 stage intensifier	IMACON 675 50/40 intensifier
Weight	75 kg	30 kg
Power consumption	2500 W	80 W
Scanspeed at film	75 mm/ns	28·5 mm/ns
Time resolution at 1·06 micron on S1 cathode	5 ps	4 ps
Total recording time	900 ps	1500 ps
Dynamic range at 10 ps exp.	16	60

Apart from the obvious advantages of size and ease of setting up on an experiment, there are a number of technical advantages.   The scan speed of the 675 is less than 40% that of the 600, yet the spatial resolution is more than proportionally greater.   Thus the system time resolution is improved for the same conditions of extraction field at the mesh grid.   The total recording time is increased from 900 to 1500 picoseconds.   With the improved time resolution, the number of resolvable elements in time across the screen is doubled, from about 150 to 300.

Due to the fibre optic coupling throughout, the full gain of the channel plate can be obtained at the film. Thus single photoelectron recording is possible on slow negative film, which has a long dynamic range. The dynamic range of the system for a 10 picosecond exposure is increased from $16 \times$ to at least $60 \times$.

Optical coupling between the sweep tube screen and intensifier cathode reduces image brightness to a lower level than at the sweep tube cathode. Thus the signal/noise ratio is primarily restricted by the low quanta at the intensifier cathode. This does not occur with fibre optic coupling and the signal-to-noise ratio is noticeably higher. In addition, the 50/40 intensifier is synchronously gated to the streak circuit, which reduces d.c. noise level accumulation to a minimum.

The systems so far described have a limiting time resolution of about 5 picoseconds when the cathode is used close to its cut-off wavelength. To obtain shorter time resolution, it is not practical just to increase extraction field strength and scan faster. The restricting factor is the image tube, as it was originally designed as a triode without the mesh grid. Figure 7 shows an equipotential plot of the P855 tube used in the Imacon 675.

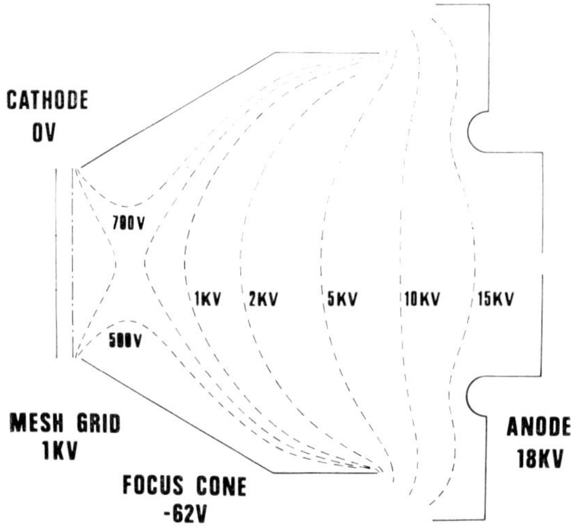

Figure 7.   Equipotential plot for 1 kV on mesh grid of English Electric P 855.

It shows the accelerating and focusing section of the tube, the deflection section being off to the right. Photoelectrons leaving the cathode are rapidly accelerated to 1 keV, pass through the mesh grid, then enter a region of reverse polarity which slows them down again. This is not significant with low voltages on the mesh grid, but as the mesh potential is increased, the reverse polarity region increases also, resulting in transit time spread occurring after the mesh grid. In addition, the spatial resolution at the screen is reduced, so the scan speed must be increased to restore the technical time resolution.

These problems have been overcome with the Photochron II image tube [2], which has redesigned cathode to anode geometry, figure 8. The mesh grid is only 0·5 mm from the cathode with 1 kV applied to it. The cathode and grid assembly are moved several cm further forward of the focus cone, which not

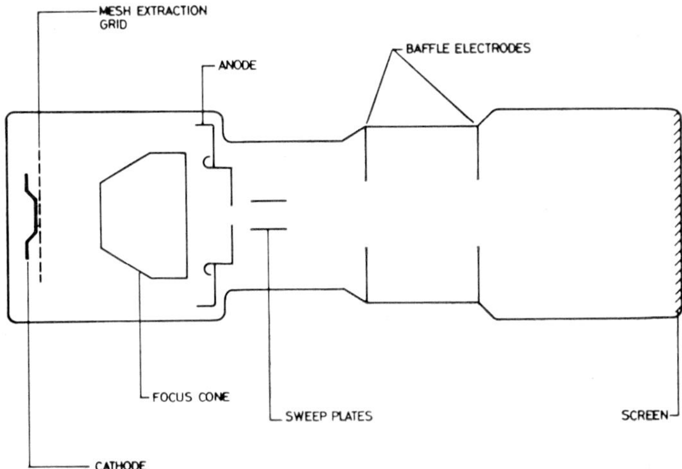

Figure 8. Photochron II image tube, redesigned to reduce transit-time spread and improve spatial resolution.

only ensures a positive field after the mesh grid but also reduces the cathode to screen magnification. This is important because the high magnification in P855 and Photochron I tubes reduces intensity/unit area, so the intensifier must be operated at higher gain levels to reach a recordable intensity level.

Table 2. Characteristics of two piscosecond image tubes.

Performance characteristics	**Photochron I**	**Photochron II**
Spatial resolution at phosphor	8 lp/mm	18 lp/mm
Spatial resolution at photocathode	30 lp/mm	36 lp/mm
Electron-optical magnification	× 3·6	× 2
Photocathode extraction field	6·6 kV cm^{-1}	20 kV cm^{-1}

A comparison of the performance of Photochron I and Photochron II tubes is shown in Table 2. The static resolution is improved by 2·25 × and when incorporated in a four stage intensifier system, the dynamic spatial resolution is effectively doubled. Thus at the same scan speed, the technical time resolution of the Photochron II is a factor of 2 × better than the Photochron I. However, this is achieved with an extraction field of 20 kV/cm compared with 6 kV/cm, so the transit time spread function is reduced by more than 3 times. As a result, at a scan speed of 200 mm/nanosecond, the Photochron II has sub-picosecond time resolution for wavelengths close to the photocathode cut-off and less than 2 picoseconds time resolution throughout the visible spectrum on S20 cathodes.

The Photochron II tube is presently only available in the ICC512 camera, but in the future it will be incorporated in the Imacon 675 with a fibre optic output screen.

The present interest in laser produced plasmas for compression and fusion, and in the development of short wavelength laser sources has generated the need for recording picosecond duration events in the x-ray and vacuum u.v. spectral regions.

X-ray streak cameras have been under development in several labs in Europe and the USA for the last few years. As a result of combined effort between Queen's University of Belfast, Imperial College of London [3] and ourselves, we have produced the first commercially available x-ray streak camera, the X-Chron 400 (figure 9).

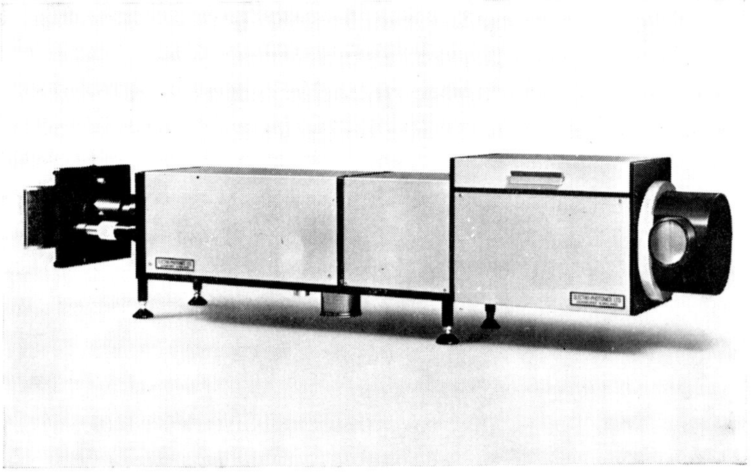

Figure 9. X-Chron, the first commercially available x-ray streak camera.

This camera is based on the ICC512 conventional streak camera but employs a gold cathode which can be changed to optimize the performance for a particular wavelength. The tube has a vacuum flange which allows it to be coupled directly to the experimental chamber thus avoiding absorption of lower energy x-rays in the air interspace. The tube is evacuated through the experimental chamber, by the normal vacuum pump.

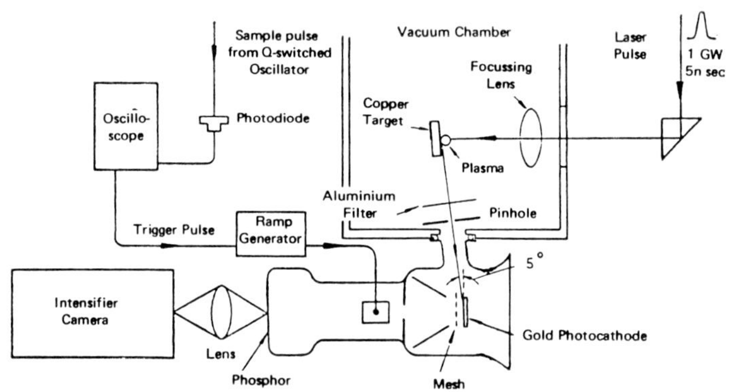

Figure 10. Recording x-radiation in streak mode.

Figure 10 shows a typical experimental arrangement in which the x-radiation from a laser produced plasma on a copper target is being recorded in streak mode. The required energy range of x-ray is selected using metal or organic film filters.

Most experimental cameras use a conventional transmission cathode, whereas the X-Chron 400 is basically designed to use a reflection cathode. This has a number of advantages, the most important being the increase in quantum efficiency. At the optimum angle of incidence, which is about 5 degrees, the quantum efficiency is higher than 30% for 1 keV photons. This is about an order higher than for transmission x-ray cathodes and becomes particularly important as x-ray pinhole optics of sufficient resolution have a very poor f-number (figure 11).

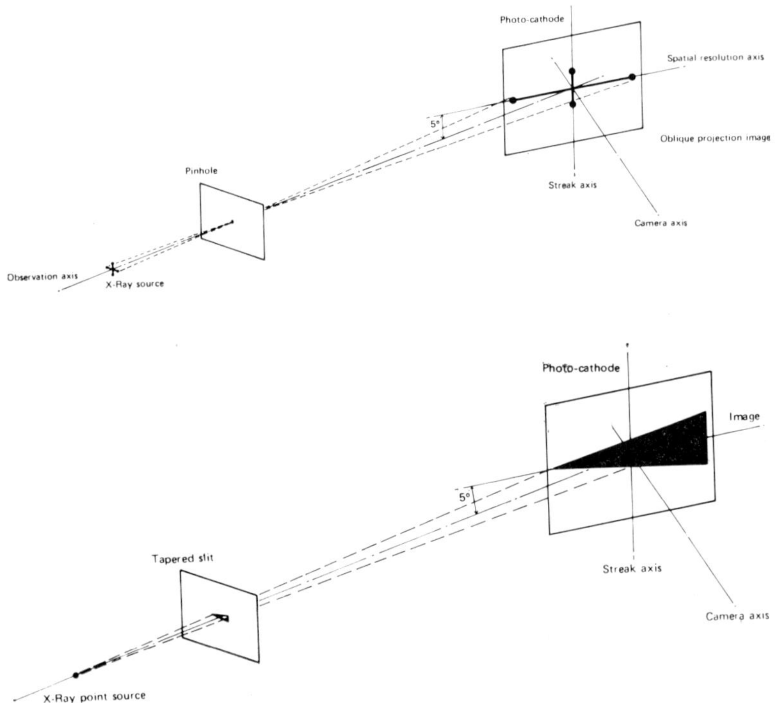

Figure 11. Simple pinhole imaging versus shadow projection of a tapered slit. The X-Chron 400 uses a reflection cathode to increase quantum efficiency.

For small plasmas where no spatial information is required, the pin hole can be replaced with a tapered slit. At 5 degrees, this produces a line image of graduated intensity perpendicular to the streak axis. Thus intensity versus time measurements can be made over a greater range than the dynamic range of the camera/intensifier system. The dynamic resolution of the system is about 5 line pairs/mm at an extraction field strength of 6 kV/cm. This sets the time resolution at about 20 picoseconds, which has been verified experimentally.

Another problem with transmission cathodes is that higher energy x-rays may penetrate the cathode and cause spurious emission from other electrodes. This is largely avoided with the reflection method, although we have found it necessary to incorporate baffles in the tube to eliminate scattered radiation entirely. However, the cathode end of the image tube is removable and this allows transmission type cathodes to be used if required.

The instruments described so far have all been designed to operate in the picosecond time region. However we have been developing new recording techniques for operating at much slower speeds.

We recently had the requirement to observe the movement of metal particles inside the air intake of a jet engine. As the particles had to be observed through the side of the engine, some combination of an x-ray source and high speed framing camera was required. In order to obtain sharp images of the particles, it was necessary to use exposures in the order of 20 microseconds and sample at 200 microsecond intervals.

A number of systems have already been developed for cine radiography, which use either a rapid burst of x-ray pulses or a continuous x-ray source. The multiple pulse x-ray method was unsuitable as the event timing could not be determined to better than 1 second and such pulse generations cannot operate for that length of time. The continuous x-ray method was also unsuitable because of the limitations caused by the phosphor decay time. We have developed a system [4] that overcomes these problems and can be used at speeds up to 10,000 frames/second with x-ray energies from 60 keV up to 12 MeV.

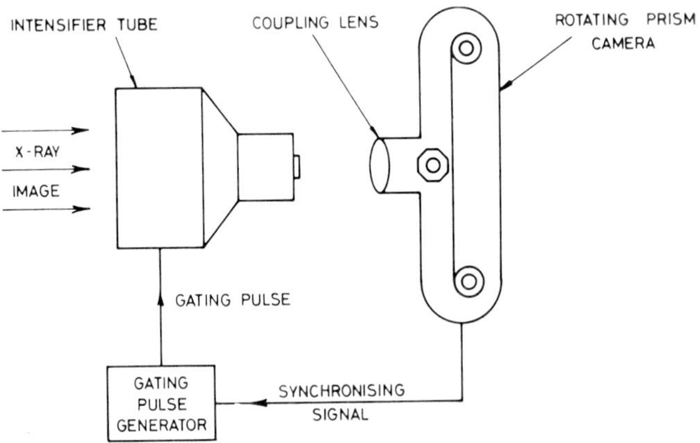

Figure 12.   System for observing movement of metal particles in a jet-engine intake.

Figure 12 shows the basic system. It consists of three main components:

1. A medical type x-ray intensifier tube;
2. A high speed rotating prism camera;
3. A gating pulse generator.

The x-ray scintillator is in direct contact with the photocathode, which avoids losses due to optical coupling. The input diameter is 300 mm, and the output diameter is 25 mm. This reduction of $12 \times$ produces a very large intensity gain in the tube, which eliminates the requirement for further intensification. This is most important because further intensification stages would increase the image decay time and so restrict the framing speed.

A Hyspeed rotating prism camera and relay lens is used to record the output screen. The Hyspeed produces synchronisation pulses at the beginning of each exposure and these are used to trigger the gating generator (figure 13). The function of the gating generator is to gate the intensifer on for a 20 microsecond period at the beginning of each exposure. This effectively freezes the image

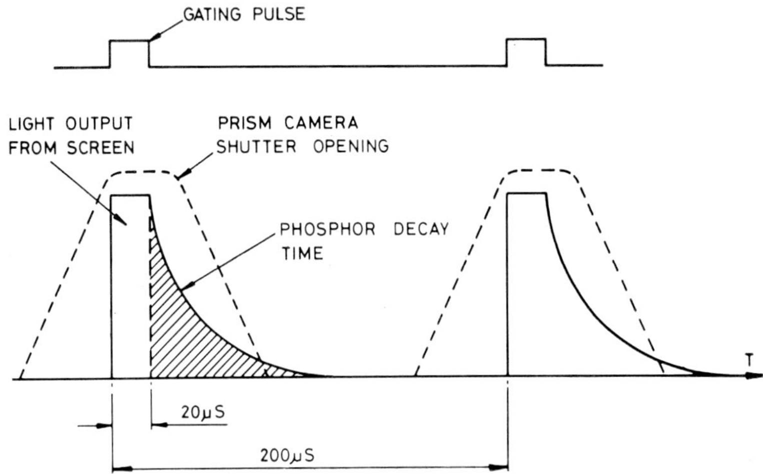

Figure 13. Hyspeed rotating-prism camera produces synchronization pulses which trigger the gating generator.

movement, which would otherwise be blurred due to the combined effect of the shutter period and phosphor decay.

Although the phosphor decay time is still about 80 microseconds with the gated method, the phosphor image is stationary and does not cause blurring.

Figure 14 shows a sequence of frames of a test subject consisting of a rotating aluminium arm with two copper discs inserted into it. An additional subject is an air gun firing a lead pellet across the frame. The disc velocities were 100 and 130 m/sec, and the air gun pellet velocity was 120 m/sec. The subject area was recorded through an 11 mm thick aluminium plate, to give equivalent attenuation to the side of the jet engine. The framing speed was 5000 frames/second and recorded on Kodak 2484 film. The intensifier input radiation level was 633 milliRads/sec at 380 keV.

Methods for recording projectiles in flight are well established and it may appear there is no need for further development. However, these techniques usually require additional lighting, which is a considerable problem with large calibre projectiles on open ranges. We have developed a gated intensifier camera [5] which can record any calibre of projectile with ambient day light, ranging from full sun to heavy overcast.

The camera consists of an interchangeable telephoto lens imaging directly onto a focused type microchannel plate intensifier (figure 15). The recording film is placed in direct contact with the fibre optic output screen. The gating circuitry generates exposures from 100 nanoseconds to 100 milliseconds and can be triggered directly from the output of a skyscreen.

Conventional recording methods usually synchronise the additional lighting, which is either by xenon flash or flash bulbs, from the projectile firing mechanism. If there is some delay in the muzzle, known as a ' hang-fire ' round, the illumination is finished before the projectile reaches the recording area. The ' hang-fire ' characteristic is of particular interest, and the gated intensifier camera can guarantee to record these rounds as it is synchronised by a skyscreen located in the recording area.

 R. Hadland

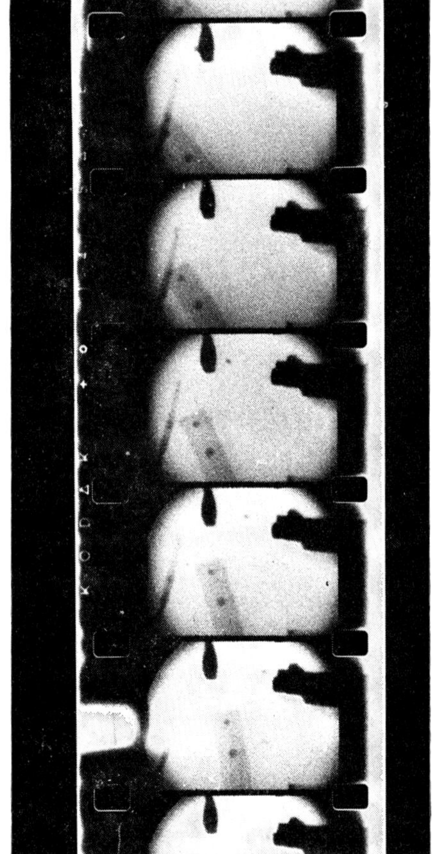

Figure 14. Test object consists of a rotating aluminium arm with two copper-disc inserts.

Figure 15. Camera for recording projectiles in flight.

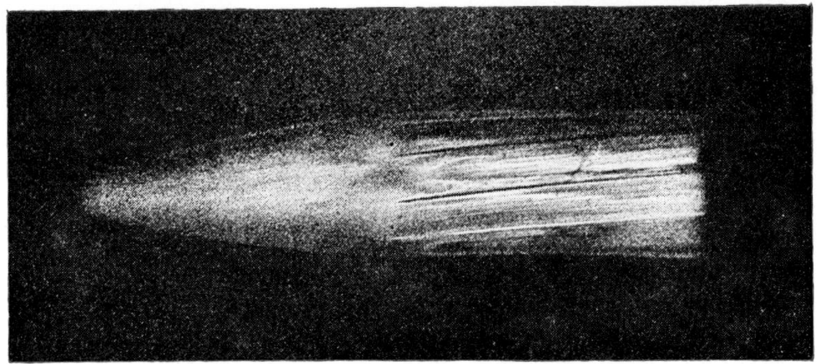

Figure 16.　Rifle bullet at 800 m/sec.

Figure 16 shows a 0·303″ rifle round with a velocity of 800 m/sec recorded indoors with standard room lighting. The exposure was 200 nanoseconds and the projectile was detected by interrupting a helium-neon laser beam.

The portability of the intensifier camera, coupled with its wide range of exposure times and S25 (near infra-red) response, make it a versatile instrument which should find a variety of other applications.

References

[1] BRADLEY, D. J., LIDDY, B., SLEAT, W. E., 1971, *Opt. Commun.*, **2**, 391.
[2] BRADLEY, D. J., NEW, G. H. C., 1974, *Proc. I.E.E.E.*, **62**, 313.
[3] BIRD, P. R., BRADLEY, D. J., RODDIE, A. C., SIBBETT, W., KEY, M. H., LAMB, M. J., LEWS, C. L. S., 1974, Proc. 11th Congress of High Speed Photography, London.
[4] HUSTON, A. E., HADLAND, JOHN (P.I.) Ltd. Internal Report, June 1975.
[5] HUSTON, A. E., HADLAND, JOHN (P.I.) Ltd. Internal Report, 1975.